WHY I WROTE THIS BOOK

This book is put together as a true field manual. Over the last 30 years I have seen hundreds of treasure related books. A few of them had more than one subject in them. For instance, one would have some good treasure stories in it. Another might have a few stories and a few pages on how to read spanish treasure symbols. Some would have how to pan for gold and some are about how to use the art of dowsing to find treasure. There are many others. You would have to buy seven or eight different books to really get started in treasure hunting!!

So that is why I have put together a field manual with all subjects needed to get started, into one book!

I also always wished that there were blank pages in the back of the book so I could make notes as I researched each story, that is now in my books.

The how-to chapters are short and basic. The goal of this book is to keep it simple , but still give plenty of information to get going and be successful.

Hopefully all the spelling is correct. Some of the grammar is not. It is that way on purpose. I wanted the words to stay exactly the way I received them from the original sources, so as to not change the original flavor and feel of the stories. A large amount of information was given to me as a small kid in the early sixties. I have no idea where some of it came from. I know some were from very old magazines and news papers. To make sure you don't get to serious, I put a few cartoons in this book that I copied when I was about ten years old, I still don't know exactly where they came from. I think they are very nifty! Any sources I could find I put in the back of the book. Some info was from interviews I did later with old railroad men and miners in some rest homes. All of these men died over twenty years ago. I misplaced some of the names over the years. I had never intended to write books on treasure hunting, I wanted to go find the gold myself!! I soon found out one man can not look for so many places, so I decided to put 30 years of info into some books so maybe some others could look for the treasures. I am a collector of stories and histories of this great state. I also collect photos of old treasure marks on trees and rocks, as well as Indian markings. If you have any of the above that you would like to share, let me know.! All things will be kept confidential. A big part of why I do it is to preserve history. A lot of times when I go back to see some old markings that I photographed, they have been destroyed by some brainless idiot shooting at them or carving their name over the old marks!!??! Logging companies have destroyed many large sections of original old growth timber. On some of these old giants of trees were names and dates from almost two hundred years ago. Now they are gone forever.

This book will take you to some of the best places in the state to find old mines and buried treasures. What a great way to spend time in the out doors! Don't be to serious. Have fun, take your time and go prepared.

Don't make or leave any messes. Always get permission to go on private land. You will have good luck and not get jinxed if you do not get greedy. Yes, I will take ten percent if this book helps you find something!!

Good luck and good hunting
David Visser

"THE STORY OF TREASURE FOUND"

Long before he reached the place where he figured he would find the treasure he was after, Frank Fish felt it was going to be a lucky day. Even he, one of America's most accomplished amateur treasure hunters, could hardly have guessed how lucky.

As his jeep bounced along a rocky trail never intended for an automotive vehicle, Fish reviewed the research that had brought him to the Guadalupe mountains of southern New Mexico. In an old abandoned mining town he hoped to find the loot that he believed two robbers had buried there back in the 1880's. One of the outlaws had disappeared, the other had been captured and died in prison. Fish's in books and other places, plus some previous digging at other likely sites had convinced him that he was now on the right track.

On the outskirts of the crumbling ghost town he zeroed in on an arroyo which, his calculations had finally told him, was the site of a shack once inhabited by the robbers. As he searched the area with his metal detector, it began signaling têe was metal underneath. After a few hours of hard digging, he had unearthed an assortment of items. On the tarp he had spread out beside the diggings, there were a number of coins dating from 1822 to 1877, a stone jug, some old locks and many keys. It was nothing like the treasure he sought, but Fish wasn't discouraged. He kept circling around the place where he had made these discoveries.

Soon his metal detector gave another high sound and he began digging again. Two feet down his shovel hit some bone fragments and, beneath them, a rusted Springfield rifle. Under the rifle, he made a gruesome find -a human skull with a bullet hole in it. Fish still did not stop digging because the detector indicated there was more metal below. Thirty inches farther down, the seeker hit his jackpot--a huge iron kettle.

Prying off the old stove lid that covered it, he beheld a mass of blackened gold and silver coins. Still more exciting was the pile of rectangular objects that lay under them. In a few minutes he had recovered eleven solid silver bars. But even that was not all. At the bottom of the kettle he found a dozen gold nuggets and two bags of gold dust!!

His time and research had paid off for him and it can for you too. This story from the early 1970's is a good example of what can happen. With today's fancy electronic doodads, it is better than ever to spend time in this great hobby.

TABLE OF CONTENTS

"No sir-not me! I am cold, wet and tired. Im not crawling through another searched out cave!"

WYOMING

STATE BY STATE TREASURE TALES

It was gold that brought the white man to where the Sioux, Cheyenne and Arapahoe had roamed for centuries in what is now Wyoming. It was gold that pitted the white man against the Indian and joined these two people into mortal combat. In the vast solitudes still lie some of the gold that men buried to retrieve at an appropriate time, a time that never came. These lost mines and hidden caches still wait for a lucky treasure hunter.

Another Lost Cabin Gold

To my knowledge, this bulletin has never been published in any treasure magazine. I quote in part from an old newspaper clipping of 1919.

"One of the most famous lost mines in Wyoming is the Lost Cabin gold mine said to be located somewhere in the Big Horn or Owl Creek mountains. In 1919 the Wymoning Historical Society published an account of the Lost Cabin mine in the Society's 'Miscellanies.' Following is the story from that publication.

"The Lost Cabin gold placers of Wyoming were discovered and worked for three days, in the fall of 1865, by seven men who came into the region from the Black Hills country. Five of the seven men were killed by the Indians; two escaped. The two who escaped brought away $7,000 in coarse gold. Since that day no effort for the discovery of this locality has been successful.

"The account given by Charley Clay, an old Wyoming pioneer, formerly of Douglas, now in Washington, March 20, 1894, is this and it is delivered directly from the two men who escaped and gave him the gold to put in the safe at the Post Trader's store at Fort Laramie:

"In October 1865, two men reached Old Fort Reno at the point which is now the crossing of Powder River, in a terribly weak and exhausted condition. They explained that they had belonged to a party of seven gold prospectors who came into the Big Horn Mountains on their eastern slope from the Black Hills of Dakota.

"They traveled along the base of the range, going south and testing the ground until they came to a park surrounded by heavy timber, through which ran a bold mountain stream, and which a few hundred yards below joined a larger stream. Here they found rich signs of the yellow metal and at a depth of three to four feet struck bedrock, where the gold was very plentiful and coarse. They immediately camped, having tools and grub which they brought by pack animals.

Among their tools they brought a big log saw and with that sawed enough logs to construct a flume. They also built a log cabin. The seven men, all working hard, finished their habitation and flume in three days and then began to work the gold in good earnest.

"Late one afternoon on the third day of gold panning they were suddenly attacked by a band of Indians and five of the men were killed almost instantly, the other two escaping to the cabin, where they held the Indians at bay until night-fall. In the darkness of the night they succeeded in escaping without being seen by the Indians. They were on foot and took nothing with them but the gold, their arms and food. From this time on they traveled at night and hid themselves during the day. After three nights of rapid and continuous walking, they reached Fort Reno and told their story.

"The two men then went to Fort Laramie and spent the winter. Here Mr. Clay met them, and being clerk of the Post Trader's store, they gave him the gold for safekeeping. He put the gold in the safe until their departure.

"They left in the spring determined to go back, and in order the better to find the place, went to the Black Hills, and formed a new party to go over their old trail. In this expedition some ten or twelve persons engaged and all were killed by Indians. For the next 12 to 15 years succeeding it was unsafe to go into that region and prospect."

I hope this information will be helpful to interested treasure hunters in Wyoming.

Kingbolt Loot

Some time in the 1800's, a lone bandit used a clever ruse to rob a stage-coach near old Fort Laramie, about three miles south of the present town of Laramie, in Albany County. The stage carried guards because it was transporting a chest containing $40,000 to meet the payroll at Fort Laramie. The bandit pretended to be sick and laid down in the road so that the stage would have to stop. Somehow, he managed to get the chest and waved the stage on. Fearing pursuit by the soldiers, he hurriedly buried the chest near the scene of the robbery and marked the spot by driving a kingbolt from a broken down wagon into the ground. The outlaw was captured a few days later and confessed to the crime. Hoping to gain last minute clemency, he told where he had buried the chest. He was then given western justice and promptly shot.

An immediate search was made for the chest. When it was not found the searchers decided that the bandit had given them false directions. In 1928, a farmer plowing along the old stagecoach route found a kingbolt driven into the ground. He knew of the buried treasure chest and is presumed to have searched for it, apparently unsuccessfully. A few years later he died without any evidence of sudden wealth.

Shaw's Ledge

In the late 1890's, Jim Shaw was with a hunting party near Mountain Home in the Medicine Bow RAnge, in the southwestern corner of Albany County. It was early fall and a light snow was on the ground. When an elk crossed the meadow ahead of the hunters and headed into the timber, Shaw told his companions to go on ahead with the wagon and make camp on the Laramie River while he stalked the elk.

Shaw soon lost the trail of his quarry and when it started snowing again he turned back to rejoin his party. He was picking his way down a steep hill when he slipped into a hole. In working his way out Shaw seized a ledge and a piece of rock broke off in his hand. He noticed at once that it was laced with wire gold. Examining the ledge further, Shaw found it to be rich in gold. He gathered all the ore he could carry and managed to make it back to the camp.

No one slept much that night. Gold fever had hit. It was agreed that the entire party would follow Shaw to the site of his find the following morning. But by morning there was a foot of snow on the ground and the snow was falling harder. The trip had to be postponed for several days. As soon as the weather cleared, the party started on their quest up the mountain. They spent ten days scouting the country for the ledge of quartz, but were never able to find it.

The next year, Shaw and a companion spent a month fruitlessly searching for the lost ledge. The following two years, Shaw made frequent and prolonged searches alone. On his last trip into the region he had an encounter with a bear and was badly mauled. This caused him to give up the search and as far as it is known, Shaw's ledge of gold is still lost.

"Big Baldy" Treasure

Bald Mountain, or Big Baldy as it is known locally, is a barren, domelike highland that rises above the black-timbered slops of the Big Horn Mountains. Big Baldy stands south of where alternate U.S. Highway 14 corsses Sheridan County into Big Horn County.

In the 1880's, Big Baldy was the center of a mining boom where fortunes were staked on finding a legendary lode of gold supposedly once worked by the Indians. So persistent were the stories of this lode that, in 1892, eastern capitalists were persuaded to install expensive mining equipment at Bald Mountain. When rumors spread that the lost lode had been found, miners rushed in by the hundreds, spawning the mushroom towns of Fortunatue and Big Baldy. It developed that the lode had not been discovered, the bubble burst, and the mining towns collapsed. The Big Baldy Lost Mine was never found.

Gold of the Sierra Madres

The Sierra Madre Mountains are a short range along the eastern slope of the Continental Divide southwest of the town of Saratoga in the south-central section of Carbon County. Indian legends have long told of rich veins of gold in this region, and rims of placer shafts and tunnels of an unknown age and origin have been found. But little gold has appeared. It is an area marked by many ghost towns, and some prospecting still goes on.

In the early days, Ed Bennett operated a ferry across Encampment Creek, about 12 miles below Saratoga. One evening a bearded stranger rapped on Bennett's caving door, then collapsed at his feet. Taking the heavy pack from the man's back, Bennett helped him to a bunk. The stranger was desperately ill with a fever. After several days his condition improved and he accepted Bennett's invitation to stay at the cabin until he was completely well. He and Bennet became quite friendly. When he was ready to leave, the man insisted upon paying his benefactor for his kindness. Bennett declined to accept anything. Still insistent, the guest went to his pack and pulled out a sample bag filled with gold dust and nuggets.

Bennett acquired from him the information that he had found the gold in the Sierra Madre Mountains which he had left before becoming snowed in for the winter, and announced that he was going east. The stranger said he would return in the spring and take Bennett to the mine. However, the man never returned. Bennett spent the rest of his active life searching for the Stranger's Lost Mine in the Sierra Madres, but was never able to locate it.

Most confident prospector I have ever seen!!

Lost "Butch Cassidy" Loot

Somewhere in the Wind River Mountains, near the town of Crowheart, George Leroy Parker, better known as Butch Cassidy, is said to have cached $70,000. Cassidy was known to have visited the vicinity frequently. At one time, he had a horse ranch near Crowheart. Discreet neighbors appeared not to notice that somehow he always sold more horses than he raised. Older residents of Crowheart say that Cassidy returned to the Wind River region to look for the buried loot in 1936, some 27 years afer his reported death in South America. Many people who knew the geniel outlaw support the belief that he did return to the United States and made a search for the loot in the Wind River region, without success. The treasure is believed to be on the Wind River Indian Reservation and permission to search must be obtained.

Sawmill Canyon Stash

Slade Canyon, also known as Sawmill Canyon, is a few miles northwest of Guernsey. The canyon is named for Joseph A. "Jack" Slade, who for a time was superintendent of the Overland Stage Line between Julesburg, Colorado, and Salt Lake City, Utah.

There is a persistent story that Slade, while with the stage line, headed a gang of outlaws who made their headquarters at Slade Canyon. The speciality of the gang was robbing emigrant trains of their stock and valuables. The stock was often sold back to the persons it was stolen from. Many searches have been made for the caches of gold, jewelry and other valuables the outlaws are said to have buried in Slade Canyon.

Slade was later hanged by vigilantes in Virginia City, Montana, for riding a horse into a general store and shooting the canned goods from the shelves.

. . . . .

The site of Fort Fetterman is on the south bank of the North Platte River, a few miles north of Douglas. Across the river and seven miles to the north of the fort was the inevitable Hog Ranch. Hog Ranch was the local name for the saloons, brothels, dancehalls and gambling houses which were the natural addition to military posts on the plains. To these places came the soldiers, cowhands, trailhands and others to spend their pay. A small cemetery usually held the bones of the slow and the careless.

When Fort Fetterman was abandoned in 1882, the Hog Ranch was acquired by Jack Saunders and Jim Lawrence. They operated it together very successfully until 1886, when Saunders was killed in a fight with Billy Bacon. It is said that Saunders regularly buried his share of the income from the Hog Ranch somewhere nearby, but it has never been reported found.

Snake River Potholes

This location of rich gold pockets, or potholes, in Wyoming, is unusual in that part of the gold was found in the 1830's, but the location was ignored for almost forty years before any more gold was taken out.

Three mountain men, after attending the trappers rendezvous in Brown's Hole, a valley where the States of Wyoming, Colorado and Utah meet, headed south to trap beaver.

Somewhere in what is now called the Grand Canyon of the Snake River, the trappers found potholes along the river, filled with gold that had been trapped when the water went down in low season. They took out a small amount, as this was before the gold rush of 1849 and gold had very little value, furs were what the trappers wanted.

One of the men, John Schuman, sent a quantity of the gold to his sister in Illinois with a letter describing the location in detail.

It is believed that the trappers were probably killed in Indian fights. Historical facts tell that there were several battles between the mountain men and Blackfeet and Gros Ventre Indians during this period.

Schuman's sister married a man named Tryell and it wasn't until about 1870 that her son became interested in the gold nuggets and old letter with directions to the site. In 1871, this son, Robert Tryell, and a friend, Norman Estell, made a trip to the Snake River. By following the old letter's directions, the two young man found the potholes.

The pair took out about forty pounds of the largest nuggets and were about ready to leave when two prospectors came down the river bank on the run, informing the young men that an outbreak of Bannocks and Shoshones Indians made it too dangerous to stay. The four men hurried on downstream. Tryell and Estell, after several days, finally reached Salt Lake City, returning to the east and home with their gold.

Tryell, after getting married, went into the mercantile business. During the minor depression of 1877-1878, he went bankrupt. Feeling sure that he could find the gold again and make a new start, Tryell again headed west. After reaching the Snake River and spending three weeks in a fruitless search, he realized that there had been so many changes that he could not find the potholes. Tryell then gave up his search and returned to Illinois.

A few years later a cousin, Elmer Nastron, was told the story of the potholes. Nastron had traveled the west for fifteen years and was familiar with prospecting. He felt he could find the location of the gold with his counsin's directions.

IN 1887, Nastron went to Lander, Wyoming, where several oldtimers told him of the three trappers who had found gold almost fifty years before in the potholes. After getting together a good outfit, Nastron traveled to the Snake River Canyon. Reaching the area that he was certain the gold was in, Nastron discovered that a huge rockslide had covered the bench above the river where the potholes were located.

After spending several days removing part of the rockslide, Nastron found the first pothole, which panned out almost $15,000 in gold dust and nuggets. The second hole produced $5,000 and the third one $2,000. After this, no more gold could be found. Nastron moved upstream and removed more of the rockslide until he finally found another pothole that yielded almost $20,000. It was then that Nastron's luck changed. A huge rock slipped and broke his leg. After crawling to his campsite and making a crude splint and crutches, Nastron waited a week before he went back to the slide.

Even with a broken leg, he removed some of the rock and discovered another pothole worth $3,000. He then decided it was time to leave. His take was about $45,000 in gold. Working his way out of the canyon, Nastron went to Idaho Falls. After selling his gold he settled in southern Utah and died there in 1940. He never returned to the Snake River and the potholes that had hid the gold. Today, somewhere in the Grand Canyon of this beautiful river, potholes with gold dust and nuggets wait for some adventurous treasure hunter.

Lance Creek Gold

On September 26, 1878, the Cheyenne-Deadwood Stagecoach, carrying between $250,000 and $400,000 in gold bullion that belonged to the Homestead Mining Company of South Dakota, was held up by five men at the Canyon Springs Station. The robbers got away with about $330,000 worth of gold ingots, currency and jewelry. After escaping and dividing the gold, the outlaws separated.

Two of the bars were found in a bank in Atlantic, Iowa. It was then learned that the banker's son, Duck Goodale, had been one of the robbers. While being taken back to Wyoming for trial, Goodale escaped.

Archie McLaughlin, one of the outlaws, supposedly buried one of the ingots near the mouth of Lance Creek in northeast Niobrara County, Wyoming. Part of the gold is supposed to be buried on Lame Johnny Creek near Sturgis, South Dakota. Another story tells that part of it was buried on Whoop Up Creek, near Newcastle, Wyoming.

Albert Spears, another gang member, was caught in Nebraska. He was given a life sentence for a murder not connected with the stage robbery. Spears told another prisoner that most of the gold was buried near the Canyon Springs Station because it was too bulky to carry.

Charles Carey was the leader of the gang. He was hanged by vigilantes in Wyoming. Frank McBride, one of the outlaws who was shot during the robbery, died while trying to get away. It was never learned what happened to his share of the gold. Another robber, named Gouch, was caught and confessed the hiding place of $110,000, which was recovered. The stage company eventually recovered all but $150,000 of the gold. The locations connected with this holdup could be worth investigating.

Rock River Treasure

On June 2, 1899, two masked men flagged a Union Pacific Railroad Passenger train at the little station of Wilcox, now an abandoned site about six miles north of the present town of Rock River. Holding their guns on the engineer, they bandits forced him to pull the express and baggage cars across the bridge beyond Wilcox and stop. They dynamited the bridge to prevent the arrival of a second section of the train, due in ten minutes, and directed the engineer to pull the cars two miles farther down the track. There the bandits looted the express car of $60,000 in unsigned bank notes.

The robbery was credited to Butch Cassidy and Ezra Lay. It is said that Lay buried his $30,000 share of the loot near the scene of the crime rather than try to pass the unsigned notes. He is supposed to have made a detailed map of the site and given it to some girl. Lay later became a respectable citizen and lived in Los Angeles.

In the 1930's, he was asked about the notes. Lay declined either to deny or confirm the story that he had recovered the currency. He is believed to have never gone back for it and that it is still buried somewhere near the bridge.

"Over here, Charlie, Im getting a ZOWIE reading on the metal detector"!!

Wyoming Brinks Stash

On January 17, 1950, a group of armed men wearing Halloween masked robbed a Brink's, Inc. office in Boston Massachusetts. The robbers took $1,218,000, the largest cash robbery in the United States history up to that time. Although seven men were later arrested, tried and convicted of the crime, only a comparatively small portion of the loot was ever recovered.

On June 15, 1962, Mack Ray Allen was arrested in San Bernardino, Calif., on a vagrancy charge. Allen told the police that in June, 1950, he had hitched a ride with an old man outside of Boston. In crossing the country, Allen said the old man had shown him two suitcases in the trunk of the car and declared that they contained $90,000 of the Brink's robbery loot. Allen further stated that when they reached Three Forks, Wyoming, he took a snub-nosed revolver from the glove compartment of the car and killed the old man. Three Forks, also known as Muddy Gap, is at the junction of State Highway 220 and U.S. 287, north of Rawlins. Allen said he hid the old man's body in a ravine near Three Forks, buried the two suitcases of money, then took the car and drove westward until he wrecked it.

Asked by the police why he had not taken the money with him, he replied that he knew the money was too hot to pass at that time. The sheriff of Carbon County returned Allen to Wyoming, but he was unable to locate the body of the old man or the buried suitcases. He was eventually released, and only Allen, who subsequently disappeared, knows whether the story he told was true or not.

Soldier's Lost Mine

In the early 1870's, General George Crook's command was encamped near Arminto in Natrona County. A soldier named Addick was given permission to hunt for antelope. When General Crook was ordered to move out, Addick was missing.

Several days later, while the command was camped at the head of Buffalo Creek, in northwestern Natrona County, Addick rejoined his company and reported that he had been lost. His time, however, had not been entirely wasted. He had stumbled upon a ledge of quartz rich in gold. In the stream at the foot of the ledge, he had picked up a pocketful of nuggets. His display of the nuggets created considerable excitement in camp. Since some of the soldiers might desert to prospect for gold, Addick was ordered to throw the nuggets away and tell no one where he had found them.

One of General Crook's scouts, J. D. Woodruff, made friends with Addick and obtained from him a fairly accurate description of the vicinity in which he had found the ledge. Soon afterward, Woodruff left his scouting job and returned to South Pass, between scouting jobs, he prospected and mined. He and a close friend, Anderson, formed a partnership with Jim Lysite, a certain Davis, and an unnamed man to search for the soldier's lost ledge. The party was ready to leave when Woodruff announced his withdrawal from the venture. Upon three successive nights, he explained, he had dreamed that while searching searching for the lost ledge all five of them had been killed by Indians. Reading a warning from the dreams, Woodruff implored the men not to make the trip. Anderson, Lysite, and the other two men were not to be deterred by what they considered Woodruff's silly Superstition. With Woodruff's description of the area to be searched, the four of them started for the ledge.

Several weeks later, O. M. Clark, known in Wyoming as "Wind River" Clark, was prospecting a small mountain in northwestern Fremont County when he noted signs of Indians in the neighborhood and hurried to depart. In the course of his flight, he came upon the bullet-riddled and mutilated bodies of four men scattered along a stream, now believed to have been Badwater Creek. The dead men were Anderson, Jim Lysite, Davis and the unnamed fourth member of the search party.

When Clark reported the death of the four men, Woodruff revealed that they had been looking for the soldier's lost ledge of gold. Several parties immediately set out from the northern section of Fremont County to find what became known as the Soldier's Lost Mine. Apparently it was never found. Lysite Mountain, Lysite Creek and the town of Lysite, all in the vicinity of Addick's supposed lost ledge, are named for Jim Lysite.

Powder House Gold

Rock Springs was about the wildest burg anywhere, including the cattle towns of Dodge City and Abilene in Kansas. The most prosperous saloon in the town was run by a Finn named Jacob Santala. It was said he had stashed away a fortune in the few short years he was there.

On July 21, 1891, Santala was behind the bar when a stranger entered. He was pleasantly surprised to find him to be a fellow Finn. The man was Jacob Hilli, from Almy. Hilli brought $500 in gold coins and was hoping to make a business deal with Santala.

After a few drinks, he got the bar owner to the side and offered his proposition. It was simple. He had $500 in gold on him and he wanted to use it to go into business with Santala. Either buy into his saloon or the two of them would start a larger one. Five hundred dollars in those days, by the way, was quite a sum of money.

In early Rock Springs, there was only one way for two parties to transact business. There had to be some hard drinking during conferences. By noon the two Finns had drunk almost a keg of beer and a couple flasks of whiskey. Somewhere during the drinking the business proposition had become lost, with the two arguing over which was the best pistol shot.

Finally, Santala declared, "Jacob, they's but one way we gonna settle this. Let's go find out."

Hilli was agreeable. It was decided that they would ride out and find a suitable target. The one to hit it first would win the other's money.

The day being hot, they decided to concentrate on some drinking as they drove out to the site of the target.

Suddenly Santala reined up. "See that building over there?" he asked, pointing toward the No. Six Town Mine.

Hilli squinted. He nodded and said, "I see it."

"Okay," said Santala, lifting the pistol and cocking it. "We'll shoot at the padlock on the door. Whoever hits it wins."

"We gonna shoot from here?" asked Hilli.

"Sure," Santala replied, trying to steady his pistol.

Santala fired, but missed the building by over ten feet. Hilli aimed, but was weaving so bad he could hardly hold the gun. Finally he squeezed the trigger. If he hit the padlock will never be known. But he had definitely hit the building and for minutes the entire town of Rock Springs trembled violently.

Jacob Santala and Jacob Hilli were never seen again in the whole. Unknown to them, their target was a powder house containing 1,213 kegs of black powder and 550 pounds of giant powder.

Though the search was painstaking after the blast, none of the gold coins that Hilli was known to be carrying were ever found.

Downey's Mine

Colonel Stephen W. Downey lived in Centennial, a small town about 50 miles west of Cheyenne during the 1870's. He reportedly worked a rich vein of gold somewhere nearby. There must be some truth to the story, for he once refused a $100,000 offer for his mine.

Eventually, Downey revealed that his vein had pinched out, and he left the area. Many prospectors refused to believe the rich vein was exhausted. Treasure hunters have searched for the remains of Downey's mine ever since.

Cheyenne River Gold

On the afternoon of June 26, 1878, the red-wheeled stagecoach raced out of Deadwood whirling a cloud of dust behind it. It was bound for Cheyenne carrying an iron box that contained $25,000 in gold.

This stage was to become one of the few successfully robbed. Its entire gold cargo was to be stolen, and $7,000 has still not been found.

Horses had been selected by the stage company for their speed and endurance. As the stage turned the bend near where the road crossed the South Fork of the Cheyenne River, near the rugged badlands, the horses started running faster, and the coach swayed dangerously as the driver leaned back, pulling hard on the reins.

A band of masked men leaped from the wall of a ravine and ran to the center of the road, shotguns raised and aimed at the men on the drivers seat.

"Stop them damn horses pronto!" one of the bandits shouted.

The driver pulled furiously on the reins and the team halted only a few feet from the robbers. During the melee that followed the driver was shot dead. The bandits climbed into the driver's seat and pushed the iron box containing the gold off the stage.

The passengers got out of the stage and stood in a line. While one bandit trained his gun on them, another went down the line filling a sack with the passengers' personal valuables and what little gold they had. Two others worked furiously, trying to open the chest with a hatchet. Seeing that they could not get it open, they decided to blast it open with gun powder. Minutes later, an explosion rent the air.

When the smoke cleared, the shattered lid revealed the sacks of gold in the iron box. But there was too much gold for the saddle bags of the robbers to hold. The rest of the sacks were distributed among the outlaw band. They then fastened the sacks to their saddles and rode off into the forest.

The guards took the coach to the next stop and reported the robbery and murder. A large posse was formed immediately and set out after the outlaws. Their tracks were easy to follow, and they were soon surrounded. They surrendered without firing a shot.

The posse counted the sacks of gold, then took the bandits baci to Deadwood, where a check disclosed that $18,000 of the gold had been recovered. But the stagecoach had carried $25,000. Where was the other $7,000 in gold?

Dilligent searches failed to uncover the gold, and to this day not a trace of it has been found. Somewhere near the Cheyenne River a historic prize awaits someone. That prize today is worth many times the $7,000 it was when it vanished.

Strawberry Canyon Treasure

It was in early spring 1870 when Grant Farley plodded into the snow-blanketed mining camp of Miners' Delight, Wyoming, after a prospecting foray into a nearby gorge.

Farley climbed off his mule and made his way slowly to the assay office, struggling through waist-high drifts with a heavy knapsack. He walked into the office and upended the sack over the assayer's desk, piling it high with big chunks of gold ore. He shrugged off his heavy fur coat and sat down in a chair, a big grin on his face.

"Tell me the value of these samples," Farley said.

Farley had discovered the ore high at one end of Strawberry Canyon, where months of winter erosion had exposed a ledge. When the assayer finished testing, it was official. Farley's chunks of gold ore were richer than any found in the region before.

After a couple drinks and a hot meal, Farley trudged back into the snow, mounted his mule and rode to the bank, where he deposited his gold ore. Then he rode back to his gold ledge and chipped out enough ore to fill his knapsack and saddlebags.

Upon his return to Miners' Delight, he was ready to file an official claim, but discovered that a commercial development company claimed title to the entire end of Strawberry Canyon. The members of the company insisted Farley show them where the gold was, but he refused. He went to the bank, exchanged his gold ore for cash, then left town heading for California.

Members of the development company and other prospectors searched for the deposit, but it was never found. The rich gold ledge still sits undisturbed in Strawberry Canyon.

On second thought, lets plant it over there.

Laramie Gold

Dewey Rascombe, 27, was the only survivor of the posse-chased bunch that had robbed the J.J. Harris bank in Laramie, Wyoming on Tuesday, May 23, 1882. They got away with $18,830 in gold coins and two bags of silver dollars, each containing one thousand uncirculated coins from the San Francisco mint.

The outlaws, four besudes Rascombe, fled into the Mummy Range southwest of Laramie. But they hadn't much more than gotten into the hills before a posse led by Marshal Gabe Hahn was on their trail. The outlaws, looking over their shoulders at the posse in the distance, dispersed into the wilderness.

Marshal Hahn was a competent lawman and he dispatched several men after each of the scattering bandits. Before the first light of dusk, the outlaws were dead, except for Dewey Rascombe, who had been thrown from his horse. He was sprawled unconscious on a grass glade when his pursuers rode up.

The posse had recovered the canvas bags of silver dollars, having shot off his horse each of the outlaws who had been carrying them. They had killed the other two, also, but neither had the poke with the bank's gold.

"He's the one got it, or at least had it," Marshal Hahn said, looking down at Rascombe's unconscious body. "We had the others in sight every dog-gone minute and they never had a chance to hide it."

But Rascombe had eluded his pursuers for a short while and, the way the lawmen put it together, during the brief time he was out of sight, he had done something with the elkhide poke that had held the bank's gold.

Rascombe was taken back to Laramie where he stood trial for the robbery, refusing to tell where he had cached the gold. His trial was held in the Laramie Public School, a one-room sawn-pine structure in which grades 1 through

8 were taught. The presiding judge was the Honorable Everett Coberleigh, a circuit court magistrate of the First Wyoming District.

The jury consisted of 12 locals, including Rolly Grenard, brother-in-law of bank Teller Ira McCloud, who had been fatally shot during the robbery.

"I didn't shoot that pore cuss," Rascombe testified, "and that's the pure gospel truth."

"If you didn't you would have if one of them other trash hadn't done it first!" Grenard bellowed.

The rest of the jury nodded their heads in agreement, they found Rascombe guilty. Grenard, the jury foreman, told Judge Coberleigh the jury recommended, "The sumbich be hung by his neck till he kicks off."

Judge Coberleigh set the execution date: "First light on Friday, June 19, 1882." This was just four days away. Rascombe dropped through the gallows trap at dawn on Friday, June 29. His neck broke with a sound like a busted stick. If the over $18,000 in gold has ever been found, it was not reported.

Lee's Placer

In the early 1860's, word sifted eastward of rich gold strikes in Montana Territory. This news attracted the attention of a steady flow of prospectors, one of whom was G. T. Lee. Lee was a man who dreamed of finding his fortune in the gold fields. Unlike most, Lee made a rich strike, only to lose it.

In 1877, Lee's story appeared in R. E. Strahorn's extensively researched HANDBOOK OF WYOMING AND GUIDE TO THE BLACK HILLS. The book contained a great deal of information on the gold rush that swept the Black Hills after gold was discovered by members of General George A Custer's exploratory expedition in 1874.

However, some prospectors, including Lee, had found gold in the Black Hills much earlier. In 1863, Lee and 12 companions set out from Missouri on the long trek to the gold fields of Montana Territory. Traveling to the north of Fort Laramie Road, the party came within sight of the Black Hills, in what is not South Dakota. The men decided to pause briefly to explore the hills before moving on.

In a stream in a deep ravine well beyond sight of the plains below, they found gold in such large amounts that they delayed their plans to go to Montana Territory. Over the following months, they steadily worked the gold-laden stream, first by simply panning it. As they found more gold, they built sluice boxes to speed the recovery.

The first winter snows forced the men to reluctantly abandon their claim, but they swore to return the next spring. Moving back down to the plains, they pushed on to Montana Territory, reaching Alder Gulch in December 1863. They soon forgot their vow to return to the Black Hills, for they found the Montana gold fields so rich that they worked them for several years.

In 1876, Lee moved to Central City. Rather than seek gold in the fields, he pursued it by opening a small shop in town. However, he had not forgotten the rich placer deposit he and his companions had worked so long before. In his spare time he traveled through the gold fields and was convinced that none of them resembled the deep ravine. He told Strahorn the rich placer had yielded so much gold was still lost. Lee's deep ravine and rich placer deposit remains lost.

Ella's Gold

Sandy-haired Ella Watson and her lover, Postmaster Jim Averill, were hanged in a gulch near the old Bothwell Ranch in Sweetwater River country of Wyoming one summer's day in 1899. They met their fate under mysterious circumstances. Whether the lynching was intentional or not is still debated among Wyoming cattlemen.

However, no one disputes that with their deaths no one remained who knew where the pair had cached $10,000 to $50,000 in gold and silver which they had accumulated from stolen cattle deals. The two had been partners for several years and at one time Averill had owned a saloon near the Bothwell ranch.

The valuable caches have lured treasure hunters over the years, but not one is known to have been uncovered. Both Averill's saloon and Ella's ranch would be good places to start looking, if you could learn exactly where they stood in Sweetwater Valley.

"Big-Nose" George's Cache

Robbery in the old west was commonplace, and hiding the loot was the only way to handle it with a blood-thirsty posse on your hills. Some of the thieves lived out their years in prison, but many of them, like "Big Nose" Geroge Perrott, with a half-a-million dollars cached away, died a violent death. He and his gang specialized in robbing gold and payroll shipments. They all died before they could spend a fraction of their loot, or so it is believed by historians.

No one knows for sure how much 'Big Nose' George managed to steal, but the stages often carried as much as $10,000 in gold each trip.

One of "Big Nose" George's caches is believed to be near Beaver Creek, Wyoming, as the gang robbed the Deadwood to Cheyenne Stage several times. He was caught and tried December 15, 1880, and was hanged by a mob.

"Big None" George came to a macabre end but his many caches of loot remain hidden in Wyoming, Montana and South Dakota, waiting for the treasure hunter to uncover them.

Two of the men were killed by Indians. Hulburt escaped with his life, but he could never find his way back to the fabulously rich—

LOST CABIN MINE

The chronicles of the Old West are liberally sprinkled with tales of buried treasures and lost mines. The Big Horn Mountains of Wyoming are the setting for many of these fascinating stories. This is the saga of one of them.

The Lost Cabin Mine is one of the most famous and elusive of all the lost mines of the west. The story of its discovery and consequent abandonment reads similar to that of another fabulous lost lode — the Lost Adams Diggings in New Mexico, which was the subject of an excellent article by Steve Wilson in the August, 1968 issue of **True Treasure**.

Many, many tales have been told of the Lost Cabin Mine in the span of years since its discovery. As is often the case, fact and fiction have become so interwoven that it is now difficult to separate them. Yet there are enough known facts concerning the Lost Cabin Mine to support the statement that it actually did — and still does — exist.

But the many stories that have sprung up about the Lost Cabin Mine make exciting reading, so it is proper that they be told here, along with the most authentic story of the lost golden lode that has been handed down to us.

The most reliable account of the

Until winter gripped the land in an icy hand, they worked from daylight to dark taking out gold, and each man realized about $100 daily.

37

discovery of this famous lode is taken from Alfred James Mekler's book, a "History of Natrona County, Wyoming," which was published by R. R. Donnelley and Sons in 1923.

Mekler quotes as his reference an article by Charles K. Bucknum of Casper published June 24, 1897. In his book, Mekler states that Bucknum came to Wyoming and Montana in the early days of those territories, and that he had participated in the gold rush to Bannock and Virginia City.

In his article, Bucknum wrote that a prospector by the name of Allen Hulburt, a forty-niner from the California gold rush days, was one of a trio who discovered the rich lode. Hulburt helped erect the cabin which gave the mine its name. But Hulburt's two partners were killed in an Indian attack, which drove him away; and although he tried, he was never able to locate the mine again.

The story of Hulburt and his discovery is contained in the following excerpt:

"Hulburt joined the stampede to the West Coast when gold was discovered in California in 1849.

He spent quite a few years on the West Coast, then headed into Oregon and wound up in Walla Walla, Washington stoney broke. That was the spring of 1863. He teamed up with two other prospectors and in some manner managed to get together a grub-stake for a prospecting trip. When the trio left Washington, they carried along six horses and a month's supply of provisions. They took the Mullen trail and headed for the Rockies. Eventually they reached the Yellowstone after a laborious journey, and down this waterway they proceeded via rafts until they reached the Big Horn River, and there they pitched camp in a veritable wilderness thickly infested with marauding Indians.

"They decided to prospect the high range of mountains just in front of them, but traveled under cover of darkness as they feared an Indian attack by day. The further they penetrated the great range, the more color they discovered. And then one day they happened upon a fabulously rich vein. There they sank a shaft and hit rich (pay) dirt at seven feet down.

"They decided to reap their fortunes from the gold they had uncovered. They had an ample supply of ammunition and game was to be had for the hunting. Even though Indians were in an abundance in the vicinity, the lure of gold was greater than fear and they decided to spend the winter panning and digging for gold. First they cut lumber and constructed a dam across the creek which was there, and put up a sluice box. Until winter gripped the land in an icy hand they worked from daylight till dark taking out gold, and each man realized about $100 daily.

"As winter made its inroads, the creek froze solid, thus preventing any more mining. The men then whipsawed enough lumber to build a cabin. As a bulwark against the hostile redmen, they threw up a stockade around the structure. There was plenty of grass so they laid in a supply of hay for the horses and dug in for the winter.

"Months later, when spring replaced the snows and frosts, the three prospectors hurried to their sluice boxes and again reaped a golden harvest. For a while, all went well, and then one day disaster struck with hurricane force in the form of a vicious Indian attack. Hulburt left the creek and went to the cabin for some reason. He was just out of sight of the other two partners when suddenly rifle shots ruptured the still mountain air. Rushing back, he saw both prospectors fall dead at the hands of the hostiles.

"In the excitement he managed to escape being seen and hastily concealed himself. Finally, the Indians left and he crept out and went to the cabin, where he hastily packed some food into his bag and, carrying some of his gold, left on foot as the Indians had taken the horses along with them. On the trail where the three men had entered the mountains, Hulburt found Indians in camp, so he turned to the south and hastened through a veritable wilderness.

"Overcome with fear, he neglected to take any notice of landmarks. He wandered for many days, sustaining himself on roots and berries and small game which he caught with a snare. After a hard journey he eventually reached the prairies.

"He struck out from that point in a direction which would reach the old Oregon Trail, which he had gone along on some 15 years previous on his way to California. After another 18 days of hardship, he came to a crossing of the Platte River at Reshaw Bridge, a distance of three miles from the site of present Casper, Wyoming. And it was there that he met his first white man on his trip.

"Since his absence from the country, gold had been discovered at Grasshopper and Adler creeks in Montana. It was at Reshaw Bridge that he ran into a stampeding mass of men heading for Montana. He related to some of his fellow prospec-

About 550 men, women and children elected to go with Hulburt in search of the gold. The party possessed about 150 wagons.

39

The Big Horn Mountains in Wyoming. Somewhere in this rough, wild region is located one of the most famous and elusive lost mines of the old west.

tors the story of the fabulous vein of gold he and his two partners had located in the Big Horns, and as proof of his story, displayed the gold he had taken when he fled the mountains. The men to whom Hulburt told his story were fired with the ambition to relocate the mine, so about 550 men, women and children elected to go with Hulburt in search of the gold. The party possessed about 150 wagons.

"For several months Hulburt led the eager searchers through the mountains, but he found no trace of his lost mine and finally was forced to admit that he was completely lost and that he did not know where to look for the mine. Some of the men with Hulburt became so furious with him that they wanted to hang him, but were halted when

one of the men of the party stopped them at gun point. The Indians were after the whites and eventually the party broke up and went their way,

"If you can stop lying about that great treasure discovery long enough, your dinner is ready."

most of them heading for Virginia City, Montana. It was in that spot that Hulburt was last seen in the autumn of 1864."

The number of stories that have been handed down concerning the Lost Cabin Mine are legion, some based on truth, and some more on fabrication.

One story tells of two miners who appeared at Fort Fetterman in the year 1865. They related that they had been members of a party of prospectors who had discovered gold in the Big Horns, had erected a cabin, and had proceeded to mine the gold until they were jumped one day by a band of hostile Indians who killed all of the miners but the two who told the story.

As proof of their tale, they

40

showed the golden nuggets they had taken which had been stored in baking powder cans. They left the garrison, saying they were heading for the east. But they were never seen again -- that is, they were never seen again by the personnel and troopers of Fort Fetterman.

It is a known fact that Thomas Paige Comstock, after whom the famous Comstock Lode was named, and who claimed to have discovered the fabulous vein of silver around Sun Mountain in Virginia City, Nevada, was backed by a group of mining men in an attempt to relocate the Lost Cabin Mine of the Big Horns.. This event took place in 1870. Comstock failed miserably, and a short time after his return from his wild goose chase in the Big Horns, he committed suicide in Bozeman, Montana.

There is another story to the effect that a very old man who was reputed to know the location of the Lost Cabin Mine showed up in Buffalo, Wyoming with three young companions determined to refind it. Unhappily, as the story goes, the old man fell dead as he was boarding a wagon, and the secret died with him.

An article concerning the Lost Cabin Mine appeared in the *Casper Tribune* of August 17, 1893. It stated that a prospector named Carter had appeared in Casper and announced that he had discovered the Lost Cabin Mine in the mountains of Wyoming. He carried with him some samples of cement rock which he said came from the tunnels located near the remains of the cabin.

According to Carter's story, he had left Montana with a party of men, one of whom was quite certain he could lead the way to the famed Lost Cabin Mine in Wyoming. After five days of fruitless searching, all of the party except Carter and one companion gave up. Carter and this man continued to look for the mine and eventually came upon a thicket. While Carter was laboriously working his way through the jungle-like growth, he found some ancient logs about two feet above ground.

Although showing the consuming effects of time and the elements, the logs had obviously once been a part

of a rude cabin. In one corner of the decaying edifice Carter found what had once been a door. The roof of branches had fallen in.

The two men were jubilant, as they were certain they had located the elusive Lost Cabin Mine. They set out to hunt for gold and a short distance from the rotting cabin came upon a series of tunnels that had partially collapsed. They took as many rock specimens from the tunnels as they could carry, and set out for Casper. There a man by the name of Lily pounded the ore in a mortar and found traces of gold.

Several Casper businessmen, interested in Carter's story, raised a fund of $100 to finance an expedition of six men to hunt the lost

Some of the men with Hulburt became so furious with him that they wanted to hang him, but were halted when one of the men of the party stopped them at gun point.

mine. The six men outfitted and returned to the Big Horns with Carter.

After three days of travel, the party reached the remains of the cabin. Their hopes were soon dashed, however, when they discovered that it was actually an old blind put up by the Indians. The Indians would conceal themselves in the blind and kill the game when it came within range.

The expedition gave up and returned to Casper empty-handed.

It was during the summer of 1897 that a man by the name of C. T. "Rattlesnake" Jones said he had happened upon the lost cabin associated with the Lost Cabin Mine.

Jones carried live rattlesnakes in his pockets, and often when in a conversation with some friend would draw out a snake and pet it. Once Jones came into town with a sack of human bones which he claimed were the remains of the prospector who had discovered the Lost Cabin Mine. He also laid claim to having been led to the site by "spirits."

The foregoing accounts are just a few of the many, many stories about the Lost Cabin Mine.

It is said that Father Jean Pierre DeSmet, the Jesuit missionary to the Indians, knew of the great deposit of gold in the Lost Cabin Mine, having learned it from his red children. It is a matter of record that he once said "I know where gold exists in the Rocky Mountains in such an abundance that, if made known, it would astonish the world."

But the good father would never tell the location of these fabulous deposits because he had promised his Indians never to reveal their secret.

It is also said that Jim Bridger, the mountain man, knew of the rich deposit, as did another trapper, La-Pondre.

Many, many men indeed have hunted for the Lost Cabin Mine. But today, more than a century after its reputed discovery, it is still as lost as ever. Although much fiction has sprung up about it, there are still sufficient substantiated facts to justify saying that it did exist, and that it may yet be rediscovered.

Only time will tell.

The Lost Mine of the Bear Lodge Mountains

A Wyoming guidebook calls Devil's Tower the most conspicuous geological feature in the entire Black Hills region. It rises about 1,200 feet above the nearby Belle Fourche River and the total height above sea level is 5,260 feet. Although this tower seems completely independent of the world around it, geologists ascribe the formation to the Black Hills uplift some 20 million years ago. It was, they say, a bubble in the up-boiling lava whose quick cooling formed prismatic columns around the sides. Most of these are pentagonal although erosion during the past several million years has extensively rounded them.

Somewhere near this majestic butte there is believed to be a lost fortune in gold — shrouded in mystery and secrecy for more than a hundred years.

The nearby Black Hills of Dakota and Wyoming are situated just to the east and south of Devil's Tower and are nestled between the Belle Fourche River and southern forks of the Cheyenne River. They rise more than 5,000 feet above the surrounding plains and their rocky slopes are covered with a thick growth of pine which, seen from the distance, fades to a dark purple and gives the area the Indian name, Paha Sapa, the Hills that are black.

Lewis and Clark referred to this entire region as the Black Mountains when they passed to the east and north of them. Later, the Laramie Range, in the corner of Wyoming, was frequently called the Black Hills and it was not until the gold rush era that the Black Hills and Bear Lodge Mountains in Wyoming were known as separate ranges.

It was in this frontier setting that the Forty-Niners followed the Oregon Trail past Wyoming. During this gold discovery era the rugged Bear Lodge Mountains were the source of several rich strikes, one a lost, but very rich, lode of gold that has defied treasure hunters for many years.

Lost treasure legends abound by the hundreds and perhaps thousands. The serious treasure hunter must always look with a discriminating eye at the potential authenticity of any lost treasure. Thorough research is always required and on this basis it is quite probable that somewhere at the bottom of a deep ravine in the rugged Bear Lodge Mountains, not far from Devil's Tower, lies the remains of a small stone cabin and a very rich ledge of gold.

To fully comprehend this lost treasure story, one must turn back the time to well over a century ago. The gold rush into the Black Hills region was a time of excitement and delightful chaos. The first rich discovery claims were made along Deadwood Creek and the word spread almost immediately throughout the world. Prospectors came by the thousands and no one knew what to believe of the rumors that spread throughout the hills.

It was during this time that a grizzled miner walked into a Deadwood, S.D., saloon with his hands shoved deep into his pockets. He was recognized by another prospector who invited him over for a drink. When asked where he'd been for the past several months, the old prospector slowly pulled from his pocket several large gleaming gold nuggets. The men in the saloon hurriedly gathered around asking questions by the hundreds. The questions came from all directions. The old prospector simply smiled, shook his head slowly and related the following story.

"There's lots of gold up there. I'd of had a whole lot more if I could find where I built my cabin." Pausing momentarily, he continued. "I built that cabin out of stone in the mouth of a gulch that has the richest diggings in the world but I just can't find it anymore."

The disclosure made by the prospector was definitely not discreet and the miners in the saloon knew it. If he was giving them a bill of goods to sell them a salted mine, then it was quite possible that he was simply leading them on. However, he stated that he did not know where his rich strike was located. That, to many, sounded like the real thing.

As they were continuing their conversation, the saloon doors swung open and in walked a local attorney named Burns. The old prospector immediately jumped up and shouted: "Mr. Burns, you was there with me. Tell 'em I'm telling the truth. There is a stone cabin and a gold vein up there in the Bear Lodge and I can't find the place. It's the God's truth, Mr. Burns. I just can't find the place again."

Burns was taken aback, and then slowly replied. "It's true. Yes, I was there with him when he was building the cabin. I saw some of the gold too. It's in broken country, a very deep ravine and you just can't see the damn thing til you're right on top of it."

The old prospector finished his drink and said, "I've walked from one end of the Bear Lodge Mountains to the other and I just can't find that cabin."

Other miners kept buying the drinks for the prospector hoping to get more information about the location of the lost gold site but to no avail. The old prospector could not even give a good landmark, other than somewhere east and south of a huge mountain — which undoubtedly is today's Devil's Tower.

Sometime later, the local newspaper in Deadwood printed the following article. "A prospector recently in from the Bear Lodge Mountains says he found a rich vein of gold. The miners now in that country are hunting for a little stone cabin in the mouth of a deep gulch which contains a rich mine. The old prospector who erected the cabin in question is now unable to find it, as it is in broken country and in a deep ravine. It can't be seen until you are almost upon it. Some party in this city has been at the cabin, namely Mr. Burns, the attorney, but it seems no one can go to it the second time. If these diggings are eventually found, and prove as rich as it is said they are — look out for a big stampede."

In 1875, Walter P. Jenney, a geologist, explored the Black Hills for the express purpose of mapping the area and investigating the rumors of gold. His maps show the Bear Lodge Mountains to be that northwestern section of the Black Hills located in Wyoming.

His records describe the region as follows: "Covering an area, mostly between the Redwater and the Belle Fourche, this region takes its name from a singularly shaped trachyte butte, 'Mato Tipi' or the 'Bear Lodge.' Surmounting a hill near the north back of the latter stream, this butte forms the most conspicuous landmark in the region, resembling the base of a ruined and crumbling column, with its shaft nearly 500 feet in height and the top 1,127 feet above the waters in the Belle Fourche."

Undoubtedly, the landmark referred to by Jenney is the Devil's Tower. In any case, this general area was known then, and now, as the Bear Lodge Mountains. Somewhere within this region is an old crumbled stone cabin and a very rich vein of gold — still waiting to be claimed.

Richard Jones

Wyoming Jade

**Don't overlook nature's treasures. That dull-looking
stone could be jade—and worth up to $200 per pound!**

Do you hunt treasure for profit or pleasure? Jade will give you both! The nephrite jade found in Wyoming can bring you from $5 to $200 per pound, depending on grade and color.

Emerald green, apple green, olive, forest green, black and mottled green are some of the colors of nephrite jade. This, in my opinion, is a treasure worth looking for. And if you're more interested in making beautiful jewelry, it's a pleasure to work with jade.

Many amateur rockhounds look for jade but give up too soon because they get confused and aren't sure what they're looking for. For

example, this story was told to me by a man who has hunted and collected jade for 30 years in Wyoming.

"Early this summer, I took my weekly walk to check my fence line along the highway," he said. "As I crossed an old drainage ditch that ran under the highway, I noticed a large rock which I knew hadn't been there the week before. Almost immediately I was sure it was jade. I lifted it to my shoulder and took it home to further test it. It turned out to be a piece of very good quality jade."

At this point in his story, I asked him how the jade got into the ditch.

"It's the same old story," he replied. "A tourist or rockhound probably picked it up in the mountains somewhere, thinking it was jade. But the farther he got down the highway, the more convinced he became that it wasn't, and simply tossed it out of his car to lighten the load. By the way, I sold that piece of jade for almost $300."

I just stood there and shook my head. This same man later told me that several times he and other fellow jade hunters had picked up pieces of jade from rock piles along the highway that probably had been dumped there by disillusioned rockhounds.

I'd like to help these unsure jade hunters. There are five tests that can be made right where the questionable rock is found.

First, pick a sunny day to hunt for jade, because you need full sunlight for the first test. Chip the rock. If the chipped area sparkles, it is **NOT** jade.

Second, if the chipped area has a moon-shaped fracture, like the type of fracture you have with agate, quartz or obsidian, it is **NOT** jade. The break should be ragged.

Third, jade should feel and look waxy. The chipped area will look waxy. Even the outside rind, which is found on most jade, has a waxy feeling and a soft, glossy shine.

Fourth, hold a chip or a thin edge up to the sun. It will be translucent and in some shade of green. Black jade is a very dark green when held up to the sun.

Fifth, jade is heavier than most

rock of the same size and CANNOT be scratched by the point of a knife.

If the rock passes all of these tests, there is a good chance that it is jade. Don't let doubts overtake you— hang onto that rock until you've made the final test at home. The last, but important, step is to apply a corner of the rock to a drum sander. It should immediately become glossy, and as the stone gets hot it should not show heat fractures.

Nephrite jade comes in two different forms, slicks and rind coated. Slicks are sand or glacier-worn and have an almost polished look. These are usually found in deserts, rivers or other streams. They sometimes have a small amount of white coating on the outside, and are usually pitted.

Rind coated jade is usually found in the mountains or in the foothills and valleys. The rind on the outside is a deteriorated jade coating about one-eighth of an inch thick. The rind could be chalky-white or brownish-red, which, at a distance, makes it hard to distinguish jade from the common white or rust-stained rocks that lie everywhere. This is where the rock hammer comes into play. Every suspicious-looking rock should be chipped.

If you haven't seen jade in its rough state, go to a rock shop and ask to see their cutting material or slicks. They will let you take it out in the sunlight to examine it. Study it carefully—or, better yet, buy a small sample to study. Sometimes, when you are discouraged or confused, it helps to have a sample of jade in your pocket to look at. In most cases your "jade fever" will return and the confusion will leave.

The hunting grounds in Wyoming extend from the Wind River Mountains in the west to the Glendo Reservoir in the east, and from the Big Horn Mountains in the north to the Colorado border in the south. Beautiful boulders of apple-green jade have been found just south of Jeffrey City in the Green Mountains.

Jade has been found in the Platte River. Around Landers is another well-known area where it is found. There is supposed to be a lost ledge of top quality jade in the Wind River Mountains. Crook's Mountain and the Red Desert are other well-known areas for the jade hunter.

But what about all of those unknown areas? I believe that there are still tons of jade to be discovered in Wyoming, and it could be waiting —-for you!—

THE GREAT TREASURE COACH ROBBERY

The trail left by the heavily-laden wagon is nearly a century old. But find it and you may find $400,000 in stolen gold!

The Cheyenne Black Hills Stage Line boasted that their specially built coach, the **Monitor**, could not be robbed. Best known as the "treasure coach" because it transported raw gold and bullion out of the Hills, it carried heavily armed guards instead of passengers.

But the jubilant stage line officials bragged too soon after the treasure coach had passed unmolested a few times over the routes between Deadwood, South Dakota, and Cheyenne, Wyoming. For then the hastily recruited Carey gang took it for every ounce of dust on board.

This robbery created a great hoard of buried gold that has not been found to this day. The getaway route was searched thoroughly at the time, and more expertly in recent years by professional treasure hunters. But no one seems to have had any luck.

The stage line never officially divulged the total amount lost to the robbers. Agents of the company estimated unofficially that the hold-up gang had gotten away with $27,000 worth of jewelry, $3,500 in currency, diamonds valued at $500, and bullion from the Homestake Mine weighing 700 pounds.

Thirty days after the robbery, the stage line announced that "three-fifths of the loss" had been recovered. However, records reveal that only two gold bricks, some dust and a part of the jewelry was recovered. What happened to the shipment of bullion?

It is still buried somewhere in the Black Hills, from all indications.

So contemptuous were stage line officials of the possibility of road agents attacking the **Monitor** that they actually set in motion the chain of events leading up to the robbery.

be worth $400,000. This did not include bullion shipments from several smaller mines.

September 26, 1878, was the red letter day for the ill-fated **Monitor**. Pulled by six horses, it wheeled into the Canyon Springs station 42 miles south of Deadwood about four o'clock in the afternoon. Gene Barnett was the driver. On the box beside him sat young Gale Hill, newly employed as a shotgun guard. Guards Scott C. Davis and Eugene S. Smith were inside, with their feet resting on nearly a ton of valuables. With them rode a company telegraph operator, H. O. Campbell, en route to his station at Jenny Stockade down the line.

Canyon Springs was not an overnight stop. It served as a change station in a lonely stretch of heavy timber. Changing teams took only about six minutes. Hosteler William Miner always had fresh horses harnessed and helpers ready to hook them in on the run. A rule for operating the treasure coach was that it never stopped anywhere for more than a few minutes.

But on this fatal afternoon, neither Miner nor anyone else was in sight when Barnett lined to a halt before the log barn. A call drew no response, so Hill carelessly descended the stage to block the rear wheels of the ponderous coach. As he did so, a hail of lead crashed around the stage from holes punched through the chinking between logs in the barn wall.

A rifle bullet dropped Hill to the ground. In an instant, the scene became one of utter panic, with lead whining everywhere. A slug through the top of the coach struck Smith's head and knocked him unconscious. That was the only way a bullet could reach those in the coach, but Davis thought Smith was dead and that the bullet that killed him had come through the steel side of the coach. The guard lost his head and leaped through the off side door, sprinting for the nearest trees.

Campbell ran after him but was shot dead in his tracks. Davis reached the trees and whirled to see five road agents charging from the barn. One of them ran to hold the leaders and keep the scared team from stampeding. He was brought down by Davis, mortally wounded.

Barnett was pulled off the stage. With a road agent behind him, he was marched toward Davis. The guard was told to drop his gun and surrender. Davis threatened to open fire, but Barnett yelled that he would be killed. Davis then tore wildly off through the timber, hastened on his way by a scattering of rifle balls.

Around nine o'clock that night the guard reached the Ben Eager ranch. Obtaining a horse, he raced on toward Beaver Creek station. But on the way he met company guards Bill Sample, Boone May and Jesse Brown. They had been waiting at Jenny Stockade to guard the stage through their division. When it was hours overdue, they became alarmed and started out to meet it.

On reaching Canyon Springs after midnight, the four guards found the looted stage pulled off the road opposit the station building. Barnett and Smith, who had regained his senses, were tied securely to the wheels. The team and some of the other horses had been driven off.

Apparently there were several other employees at Canyon Springs at the time of the robbery. Mention was made in newspaper reports that they were found tied to trees. This was done when the outlaws captured the station without firing a shot. However, hosteler Miner had been locked in the granary instead of being tied up. Making his escape during the shooting, he reached Cold Spring ranch on the north road. Grabbing a horse there, he hurried on into Deadwood to give the alarm.

The stage line was stunned to learn that their vaunted treasure coach was anything but robbery-proof. Especially so was the company superintendent, Luke Voorhees. He had a hand in designing and constructing the **Monitor**, which was specially built in Cheyenne. Heavier and larger than a Concord, the running gear and thoroughbraces were reinforced to carry a full ton of valuable cargo. The interior, except overhead, was lined with steel to turn bullets. Slanted portholes in the side doors enabled guards to fire at holdup men without danger to themselves.

An unusual feature of the treasure coach was a green painted steel box bolted to the floor. The sides and ends were three inches thick. The Cincinatti manufacturer warranted that when fitted with their special lock, this safety box, called the "Salamander," could not be opened in less than six days.

But the Carey gang made the Salamander's impregnability a mockery. According to witnesses Barnett and Smith, who were tied to the coach wheels, the road agents broke the box apart in an hour and a half! Most of the Homestake bullion was carried in it. Other valuables were transported in heavy canvas sacks. Loading all of the loot on eight horses, the road agents vanished into the night.

Not until Miner returned to Canyon Springs from Deadwood did investigating guards obtain any worthwhile clues. The hosteler reported that an hour before the scheduled arrival of the treasure coach, a lone rider had jogged in. Dismounting, he had suddenly covered Miner with a six-shooter. Only then did Miner recognize him as a man previously pointed out to him as Charles Carey.

Carey was not masked and neither were the others who rode out of the timber when Carey whistled. Working fast, the road agents tied up or locked up all those at the station. They next punched holes through chinking in the barn wall to fire through. Miner could tell the agents little more, since he had freed himself and ran for help during the short gun battle.

Charles Carey was 27 years old, stood six feet tall, was fair-complexioned and had light brown hair and a mustache. Long suspected of running with the tough outlaw gangs infesting the Black Hills, he had finally struck out on his own. The Canyon Springs holdup was his first job as a gang leader.

His was the largest haul ever made by road agents in the gold-rich Hills of Wyoming and South Dakota. It was also one of the biggest stagecoach robberies in the entire west.

The astounding news of the holdup set off a reign of excited confusion. Many wild rumors were enlarged upon and accepted as fact. The region of the Black Hills had long been plagued with murder and robbery, with more than a dozen gangs of wanton killers raiding everywhere. Appeals were now made to Fort Laramie for troops. But the Canyon Springs road agents had already made good their escape—at least, for a while.

They were quickly pursued, however. More than a dozen large bodies of heavily-armed men from Cheyenne, Deadwood, Rapid City, Custer and elsewhere dashed into the Hills —some of them hoping to earn the $2,500 reward by Superintendent Voorhees for the gang's capture.

Canyon Springs, also called Whiskey Gap, lay across the territorial border of the Dakotas in Wyoming. Here on the edge of the fantastic Black Hills, the Carey gang had a hundred choices of isolated wilderness in which to hide. As the search

for them mounted in intensity, they seemed to have vanished into nowhere.

The armed posses on their trail took into custody all suspected outlaws—and strangers. Many horse thieves and badmen of lesser fame fell into the law's net. An organization of impromptu vigilantes calling themselves "Minute Men" also got into the act, and several subsequent hangings in the Black Hills area were charged to them. But the dual hanging that was to confound the stage line's attempts to recover the Homestake loot was done by erstwhile lawmen on the trail of the Carey gang.

After dashing about futilely for more than a week north of Canyon Springs, the squad of stage company guards under the command of Scott C. Davis turned south. En route to the Jenny Stockade station, on a fork of Beaver Creek, they surprised and captured two young men who had camped for the night.

Both men sported small mustaches. One was light-complexioned, and had pale brown hair and blue eyes. Davis excused his action later by saying he did not realize the second man fitted Carey's description, for he had not then heard it from Miner.

At any rate, the young men were questioned. Their replies, while not belligerent, were of the go-to-hell variety. They denied any knowledge of the robbery, insisting they had just entered the Black Hills. They claimed to be Californians, pointing to their brass-studded trousers as proof. These were made only in San Francisco, especially for miners.

Davis was to say later their answers were unsatisfactory and that he figured them to be horse thieves and outlaws. So why waste time on their ilk?

Both men were hanged from one tree.

From Jenny Stockade Davis again led his guards to the north. But their search to Inyan Kara and east to Rawhide Butte netted them exactly nothing. Behind them, swinging to a tree, was the biggest mistake of the unfolding mystery of a huge treasure trove still lost to this day.

Meanwhile, two minor employees of the stage line, John B. Brown and Charles H. Borris, were arrested as suspects. Minute Men hauled them up on ropes to a tree limb a few times, and they confessed. For some reason the vigilantes did not hang them, but gave them over to the law.

Removed to jail in Cheyenne, both men repudiated their confessions and vigorously protested their innocence. Thereupon Brown was released, but Borris was discovered to be an escaped convict and returned to prison in Wyoming.

William Mansfield and Archibald McLaughlin were arrested on purported evidence. They couldn't give a very good account of themselves. In Deadwood they had tried to sell about 20 ounces of raw gold. A mob took over and rope torture elicited confessions. It was claimed they admitted being members of the Carey gan.

In the middle of October they were sent to Cheyenne in custody of guards Jesse Brown and James May. Supposedly, they were to await trial there. But both now denied they had confessed to anything. For some reason, they were started back to Deadwood with the same guards.

Just beyond Fort Laramie a masked party of five Minute Men halted the coach on Little Cottonwood Creek. The vaunted fighting ability of guards James and May deserted them instantly. The coach was driven under a tree and Mansfield and McLaughlin were boosted to the top. Nooses were placed about their necks and the stagecoach was driven from under them, leaving them dangling in the air.

Another suspect, Tom Price, was apprehended in Lead, South Dakota, after a desperate fight in which he was wounded. He was sent to federal prison in Lincoln, Nebraska, for another crime, but he denied being a member of the Carey gang. Later events proved he was telling the truth.

About this time, Superintendent Voorhees received a tip that Albert Spears, alias Al Spurs, was one of the road agents. This information was turned over to U.S. Marshal M. F. Leach. Trailing Spears, Leach found that he had sold $800 worth of gold dust and $500 worth of jewelry at Ogallala, Nebraska. Tracing him on to Grand Island, he actually caught Spears in the act of trying to sell other minor items of loot from the Canyon Springs robbery. The clincher was that he had Hill's gun, taken from the wounded guard at the hold-up scene.

Spears readily admitted his guilt, naming Carey as the leader. The wild bunch, he said, had been hastily gathered when it was learned how much the treasure coach would be carrying. He claimed not to have known all of the gang, or how many had actually participated in the crime. He also swore the others had done him out of his just share of the loot, and he had received only "trash" pickings.

For the first time, it was revealed that the road agents had encountered serious difficulty trying to get so much gold out of the Black Hills. It had been cached, said Spears, when the pack horses gave out under the heavy load. He told several impossible stories as to where the bullion was buried. Undoubtedly, he hoped to return some day and recover it himself.

Pleading guilty to the second degree murder of H. O. Campbell, the stage line telegraph operator killed during the robbery, he was given a life sentence. A few years later he died in the federal penitentiary at Lincoln.

Still another culprit, but one who was never proven to have participated in the treasure coach robbery, was Charles Ross, alias Jack Campgell, alias Jim Patrick. Long after the hue and cry had died down, he was arrested in Nevada and returned to Wyoming. Convicted in another robbery case, he received a sentence of 12 years at Lincoln, dying there in 1885.

The stage line's division agent, Captain W. M. Ward, led a band of manhunters into the Harney Peak area—and hit upon a trail that has been luring treasure hunters ever since. They found a rancher who had sold a light wagon and a team of horses to two men, believed to be road agents, for $250.

The posse picked up the trail of the wagon, following it to Horsehead Crossing, a station managed by George C. Boland. Boland reported that three men had arrived in the wagon, one of them badly wounded. The wounded man was 24-year-old Frank McBride.

McBride's companions drove away at once in the wagon. Boland said it was heavily loaded, with the contents covered by a tarp. McBride died that same night of gunshot wounds in the chest and right side. Boland buried him the next morning. McBride did not talk, but after hearing of the robbery Boland at once suspected what had been in the heavily-laden wagon.

Belatedly hot on the trace, the posse took off after the wagon, certain it was being used to haul the Homestake bullion. But it had a lead of several days and at times they lost the trail. Fortunately, freighters were encountered who reported seeing the wagon driven by two hardcases. It was making north, ostensibly for the Deadwood Pierre trail out of the Black Hills.

The wagon then swung east, circling Rapid City, and crossed the divide and took up Boxelder Creek. Ward's forces were now joined by a Rapid City posse under Sheriff Frank Moulton. With it was a Dr. Whitfield, destined to make an important discovery relating to the case. After sunset the combined party arrived at Washita Spring Road House, 20 miles east of Rapid City. There it put up for the night.

Resuming pursuit the next morning, the wagon was found abandoned at Pino Springs, well off the regular freight road. It was empty. Presumably, the horses had been ridden off bareback by the road agents. The posse continued on their trail but Dr. Whitfield's horse, an Indian pony, gave out and he could not keep up. He turned back for Rapid City. Down the trail below Pino Springs, he halted his pony to rest. There he found the dead ashes of a small campfire. Wheel marks indicated the fleeing road agents had camped there a part of the night.

Poking about in the ashes, Dr. Whitfield uncovered the corner of a buried garment. It proved to be a pair of blue jeans—wrapped around a gold brick of large dimensions. The ashes had been used to camouflage the cache. Excitedly, the doctor dug all about the area with a sharpened stick. But nothing more came to light.

The brick, when turned in, was worth $32,000. The doctor received a reward of $11,000.

Losing the trail of the road agents completely, the posse turned back to Rapid City. But Captain Ward continued east to Pierre, on the Missouri River. In the future South Dakota capital he struck pay dirt.

A young man answering the description of one of the road agents in the wagon had taken a train from Pierre to Atlantic, Iowa. Ward arrived in that city with nothing more to go on than a general description of the man. This stymied him, and it appeared he had reached the end of the trail. But one morning he happened to stroll into a bank owned by Almond Goodale, the most influential citizen in that section of Iowa. Boldly on exhibit in the bank was a gold brick valued at $4,300.

Ward gaped at the brick in amazement, for on the surface was stamped "Homestake #12." It was one of the consecutively numbered gold bricks taken in the Canyon Springs robbery!

Discreet inquiry disclosed that it had been brought back from the Black Hills by the banker's 22-year-old son, Thomas J. Goodale. From the west, young Goodale had sent glowing reports on how well he was doing in the mining business. Returning home, he told his father and friends that he had discovered a very rich mine, and the gold bar was received as partial payment when he sold it.

Ward had a hard time convincing local lawmen to arrest young Goodale. But when he was finally taken into custody, he had in his possession a gold watch, diamond ring, and two plain gold rings easily identified as part of the treasure coach loot.

The elder Goodale, as mentioned, was a powerful and wealthy man, and it proved difficult for Ward to secure extradition papers for his son. Finally, he did, and started back for Wyoming with young Goodale wearing riveted leg irons. At Lone Tree, Nebraska, the prisoner escaped from the train under mysterious circumstances.

How Goodale got out of the leg irons, which were found in the wash room, could never be explained. Apparently Ward had grown careless in his handling of the impressive, friendly young man who seemed resigned to his fate. At any rate, he allowed his prisoner too much freedom—probably thinking the leg irons would hold him secure; and he did not even miss Goodale until the train arrived in Ogallala.

Ward had long been a trusted agent of the stage line. But his own company suspected him of complicity in Goodale's escape. Unfounded rumor claimed that he had been paid $3,000.

Ward was discharged and a $700 reward was posted for Goodale's capture. But the young road agent had disappeared completely, never to be heard of again.

For dubious reasons, Abe Gooch was suspected of being a member of the Carey gang. Arrested at

51

Fort Thompson and taken to Rapid City, he made a deal with lawmen and led them to a spot near Pino Springs. Gold amalgam worth $11,000 was dug up. At first it was said to be a part of the Canyon Springs loot, but no amalgam had been aboard the treasure coach.

By the end of 1878, hundreds of men had searched extensively for the missing major portion of the loot. The man who drove the wagon carrying the Homestake gold was eventually identified as "Red Cloud." He was known by no other name, and was called that because he had once worked for the government at the Red Cloud Indian Agency. When sought by lawmen, it was learned that he was one of the two men hanged north of Fort Laramie by Davis and his band of stage company guards.

When it was far too late, officers recalled the incident which at the time had seemed of so little importance—the hanging of two surly hardcases. Found ten days after the execution, the bodies were so badly decomposed as to prevent identification. But if one was Red Cloud, the other was undoubtedly Carey, the gang leader. Both had vanished from the day of the hanging, never to be seen again.

The route of the wagon from Horsehead west of Buffalo Gap and east around Rapid City to where it was found abandoned at Pino Springs became the treasure trail. Surely the golden loot was buried somewhere along this known course!

At least, that was what searchers believed then—and today. The bullion was cached and the survivors took off in different directions. None was ever able to return, or so it seemed. So the gold **must** still be there—waiting!

Officials of the stage line thought so. The company placed spies and watchers all along the treasure trail. They remained there for five years, during which time both local citizens and organized groups slipped in to try finding the buried loot on their own. This action seemed to place the value of the stolen bullion quite high—perhaps as much as $400,000 as asserted.

The stage company was satisfied that four of the road agents had been accounted for—McBride, Spears, Carey and Red Cloud. But Goodale was still at large and it was thought he might sneak back and attempt to recover the loot. For this reason, some watchers were kept scattered through the country until about 1885.

Goodale's family was approached with offers of amnesty, if he would surrender and lead company agents to the cache. This again tends to prove the loss was great, but nothing ever came of the move.

For years after the big robbery, all sellers and shippers of any quantity of gold from the Black Hills were checked and accounted for. Stage line detectives continued their work on the case. But when the railroads came through, the stage line went out of business. Dissolved, it no longer cared where the bullion might be. In fact, the treasure coach robbery was forgotten for a good many years, until historians revived the story.

Treasure hunters took notice. Yearly, they slip into the Black Hills. Working from old maps, they search along the treasure trail with detectors.

There is no evidence to be found on the ground today to indicate where the wagon loaded with stolen bullion passed. The distance from the scene of the robbery at Canyon Springs to Horsehead Creek crossing is about 70 miles. The old stage station stood somewhere near the present Hot Springs airport, which in turn is about 15 miles south of Buffalo Gap.

After leaving the dying McBride at Horsehead, the remaining road agents drove the wagon north, passing west of Buffalo Gap. Their wandering route took them east of Rapid City. Apparently they intended going on to the Deadwood-Pierre trail, but for some reason they abandoned the wagon at Pino Springs.

So there you have the approximate treasure trail. The trick for hunters is to establish the route exactly, for somewhere on or barely off this meandering trail is buried an enormous sum of gold

MITCHELL AND HIS GANG
ROBBED THE WYOMING BANK
WITH NO PROBLEMS, BUT
AFTERWARD EVERYTHING WENT
WRONG

FIND WYOMING'S $68,000 LOST OUTLAW LOOT

There wasn't nothin', Cornelius (Corny) Mitchell told the six men of his sorry little gang, that could go wrong during their robbery of the Platte Depository & Loan Company in Saratoga, a southeastern Wyoming mining camp just east of the Continental Divide.

He had everything figured. "And all we gotta do afterwards is high-tail to that snakebelly land on the near side of the river and stay hid in them junipers until ever'body quits looking for us."

There was no end of grass and water for their mounts on the lush green banks of the Platte River and since Mitchell had already cached enough food for a three-week hide-out, nobody had a single thing to worry about. He was a man who thought of everything, he assured the gang. "Which is why I included a couple decks of cards with them eatables and also some fish hooks and lines in case we want to spend some of our hidey-out time fishing for trout."

Mitchell, who was 42, black-whiskered, beer-bellied and a stranger to soap and wash water, was one of the West's most insignificant gang leaders

He had been a horse trader in Clinton County, Missouri, until the War Between the States put him out of business. He enlisted in the 1st Missouri Dragoons and after the war, like many another itchy-foot veteran, he "went West," giving the Colorado gold fields a lick, whiskey running the Southern Cheyennes, faro dealing in the Wyoming trail towns and peddling wet brand beef to the Fort Laramie commissary.

But some of this required a little work now and then and Mitchell began to look for an easy road to prosperity. The route, he decided, was outlawry, so he organized a gang in the spring of 1878, recruiting his men in cheap whiskey saloons in Denver and Cheyenne and the east-slope settlements in between.

The Mitchell crowd stole horses and mules from Platte valley hay-shakers, ran whiskey and guns to the Brule Sioux and dogged Mahoney-Stanton stages on their long, rough-country runs from the Wyoming gold towns — South Pass City and Atlantic City — to Rock Springs, a Union Pacific depot town.

But none of it made the gang rich "What we should ought to do," Mitchell told his men, "is rob us a bank."

"Actually all there is to it is figuring ever'thing out first."

Mitchell and his lieutenant, Clay McFord, 31, forked their mounts into Saratoga. They sized up the Platte Depository & Loan Company, the local law enforcement which consisted of red-whiskered Marshal Gabe Holcomb, and the fastest ways to the wilderness along the Platte.

Mitchell had thought of every-thing except the possibility of someone coming upon the gang's cache of food, playing cards and fishing equipment while no one was there.

The chances of this happening were like drawing to an inside straight, but the morning of Friday, June 9, 1882, while Mitchell and his men were riding toward Saratoga to rob its bank, rancher Nate Woodrow was searching the Platte's glades and bottomland forests for mavericks that had strayed from his Logan's Butte grassland.

He found the cache. "I knew right away them victuals and devil's pasteboards and fishhooks had been hid by a middling-size bunch who were fixing to be here for quite a spell," Woodrow is quoted in *The Wyoming High Country Towns* (John H. Barry Co., 1913).

Since it wasn't likely that anyone but outlaws would be lurking along the Platte, Woodrow hightailed into Saratoga to tell Marshal Holcomb about his discovery.

Holcomb was organizing a posse. "A bunch of outlaws just robbed our bank and .44'd the teller," he explained, "so I ain't got time right now to do any socializin'."

"I think I know where they went," Woodrow said.

He told Holcomb about the cache. "That," Holcomb said, a grin on his leathery face, "is the goodest news I've heard in a coon's age."

With Woodrow on the front horse, Marshal Holcomb and the men of the posse rode toward the cache.

"It was like shooting ducks on a pond," Woodrow was to tell Florian Moore, author of *The Wyoming High Country Towns:* "We snuck up on them buggers and before they knew what was going on they were on the way to their Maker except for the one who had been rubbing down his mount, the same letting no grass grow while he raised his hands, the others having drawed and paying for it with their lives."

The posse's prisoner was Clay McFord. "You got just two seconds to fork over the $68,000 you scamps stole," Marshal Holcomb said, his 44's 3... ' end pointed at McFord's high guts, "or you are gonna be fuller of holes than a sieve."

"I forgot where we hid it," McFord said.

"Then two seconds is just about up!" the marshal bellowed.

"You ain't gonna shoot me," McFord said, "on account of then you won't never find that money."

Marshal Holcomb looked at the others. "He's got something there," he said, holstering. "We might as well take the squeaky little scamp back to town."

Circuit Judge Zachary Scott conducted McFord's trial the morning of Thursday, August 3, 1882 in the Wyoming Beer & Whiskey Emporium, the largest building in Saratoga. The charges: Robbery of the Platte Depository & Loan Co. and the fatal shooting of Rollie Childers, the bank's teller.

"I wasn't the one killed that dude," McFord said to Judge Scott. "I already told you it was Mitchell that .44'd him."

"It coulda been you if you'd been in the right place," Judge Scott said, shifting his chewing tobacco to the other side of his mouth, "so the charge stands."

A six-man jury found McFord guilty of both charges and Scott sentenced him to be hanged "until your soul departs your sinful body."

"Any use talkin' a deal?" McFord asked the judge.

"Not a particle," the judge said.

Marshal Holcomb took McFord back to the Saratoga jail followed by the bank's owner, Calvin Knox, and a half dozen men who had been depositors in the robbed bank.

"Since you're gonna be the principal at a necktie party tomorrow morning," Knox said after Holcomb put McFord in his cell, "you might as well say where you and them other rascals cached the money because it ain't going to do you a nickel's worth of good anyways."

"The onliest deal I got," McFord said, "is the one I was gonna tell that rope-brain judge. Which is let me go free and you get half the goodies."

"Who gets the other half?" Knox demanded.

"Me," McFord said, "Who the hell did you think?"

Knox said the little prisoner had more guts than a government mule. "Mister McFord, maybe you ain't noticed but you are not in a position to be making deals. Especially one in which you can steal half my money."

"It's already stole," McFord said. "My way you get half of it back. The other way you don't even get a smell of it."

At this point Marshal Holcomb said, "Why don't we go over yonder and get some beer and talk on it?"

They went across the street to the American Saloon and sat around a table and talked the situation. "There's two ways to go, the way I figure it," Marshal Holcomb said. "The first is, you gents can let me string him up and then look for the cache, which just might be like looking for a needle in a haystack. Or you can buy his deal, figuring half is a whole lot more than nothing."

Of course, the marshal added, he would expect a donation for risking his job by releasing his prisoner. "I was thinking of a little ol' ten percent, is all. Meaning ten percent of the whole thing, including his half."

"If you ain't a fine marshal," Knox said, his lips tight. "You ain't got any more principles than that little bugger over there in the jail."

"You ain't exactly a saint yourself, you money-grubbin' scudder!" the marshal said. "And you'd best think on the fact that the minute I drop the trap on that slippery little dude you won't never see that money."

The marshal and the others talked more on it before they went back to the jail. "We're buyin' your deal," the marshal said to his prisoner.

"I get half the money?" McFord asked.

"You get half," the marshal said.

"How do I know when I say there it is, you won't shoot me?" McFord asked.

The marshal said he wouldn't even think of such a sneaky thing. "I've been crick-dipped and born again and us born again folks keeps our word or we got to face eternal punishment."

McFord led the marshal and the others into the rugged area between Saratoga and the site of Mitchell's food cache. He pulled up his mount near a large shelf boulder. "There it is," he said, "in that crack in the middle, there. The one goes up and down."

A greedy smile came over Knox's face. He whipped out his .44 and fired twice into McFord's right flank.

The terribly wounded man slid off his mount. "You dumb stupid double-crossin' — —" he moaned. "This ain't the place. I was just testin' you."

He didn't have anything to lose, he said, his voice becoming weak. "I was gonna be hung anyways."

"Why, damn you!" Knox shouted, getting off his mount and running over to the dying man. "Tell me where that money actually is!"

"Ask me in hell," McFord said, barely audibly.

McFord tremored and died and Marshal Holcomb, still on his mount, said, "Mister Knox, that was the most stupidest play I ever in my whole life seen. Why didn't you wait 'till we at least found out if he was eucherin' us?"

Now, the marshal added, his lips tight, they most likely would never find the cache.

"Maybe this actually is the place," Knox blubbered. "I'm gonna find out!"

He darted to the shelf boulder and pulled himself up onto it and looked into the vertical crevice.

The money wasn't there.

Knox and the others searched the area and the region for miles around. So have others in subsequent years.

But it is a rocky wilderness with countless places in which the Mitchell crowd could have cached their $68,000 loot.

It is still there . . . a fortune for the lucky man who finds it. .

The Lost Cabin Gold Mine

The Wind River Range in western Wyoming boasts some of the most rugged country in the state. Snow-capped mountain peaks in excess of twelve thousand feet high are not uncommon, and they overlook dozens of deep, remote canyons, many of which remain relatively unexplored to this day.

In 1842, six years before the great strike at Sutter's Mill in California, gold was discovered in the Wind River Range, gold which lured several hardy and adventurous souls into this remote area which was home to hundreds of Indians. Some few found wealth, many more found only disappointment, and several found death. One of the gold discoveries in the Wind River Range may have been as rich as the greatest of the California strikes, but its location was lost more than a hundred years ago and has never been found. The gold, which apparently exists in impressive quantities, is still there.

The first gold discovered in the Wind River Range in 1842 was found by a trapper. Georgia Tom McKeever came west several years earlier to try to earn his fortune trapping beaver and selling the pelts to hat manufacturers in St. Louis, but McKeever found little but bad luck during his stay in the Rocky Mountains. Originally a member of a party of seventeen trappers, McKeever worked hard and added to the rapidly growing accumulation of pelts which the men believed would make them all rich. Three of the trappers, however, had other plans. While the rest of the group were setting traps in the ponds and streams in the Absaroka Range, the three loaded up the accumulated furs and made off with them.

Disheartened, the remaining trappers nevertheless continued their efforts for another three months until they eventually put together a second impressive collection of beaver pelts. Just as they were preparing to transport the skins to St. Louis, their camp was attacked by Indians, and all but Georgia Tom McKeever were killed. McKeever pretended to be dead, then crawled away from the camp in the dark of night and spent the next three weeks hiding in the mountains and trying to find his way back to civilization. Eventually he reached a settlement where he recovered from his ordeal.

Next season, McKeever decided to try trapping one more time. Alone, he rode into the Wind River Range to lay traps along the numerous streams. This range was normally avoided by most trappers because of the presence of hostile Indians, but that did not deter McKeever.

McKeever found the trapping in the Wind River Range disappointing. Though he diligently ran his lines and checked them every day, his take was pitifully small. One day, while searching for a new location in which to set traps, McKeever decided abandon trapping as a way of life and return to farming in Georgia. As he was exploring a new canyon, he saw a flash of color in the bottom of a narrow, shallow stream. Closer examination revealed the presence of gold nuggets in large quantities in the gravel bottom and, using only his hands, the trapper retrieved several of them.

For nearly three weeks, McKeever panned gold from the narrow stream until he ran out of containers in which to carry it. As winter was setting in, he decided to take what he had, leave the mountains, and return to Georgia.

Many days later, McKeever stopped at Fort Laramie in southeastern Wyoming and exchanged two large nuggets for cash. When asked where he got the gold, the trapper provided directions to the rich stream in the Wind River Range. Several of the hangers-on around the fort became excited about the prospect of finding gold in the western mountains, and the next morning saw nearly twenty men with pack animals loaded down with mining equipment setting out from the fort. Because of the Indian threat, however, the soldiers ordered them to return.

In 1863, two of the would-be prospectors from Fort Laramie were drinking beer in a tavern in Walla Walla, Washington, when they met a man named Allan Hurlburt. Hurlburt had arrived in California in 1849 to find gold, but had had no luck. He wandered northward over the years, prospecting and working at odd jobs along the way, until he arrived in southeastern Washington.

One evening, while taking a meal at the tavern, Hurlburt overheard the two prospectors relating the tale of Georgia Tom McKeever. Intrigued, Hurlburt introduced himself to the two men and pumped them for more information about the gold in the Wind River Range.

The next day, Hurlburt and two friends named Freitag and Smith left Walla Walla for the Wind River Range. They led six pack horses loaded with mining tools and provisions.

For weeks the three men roamed the mountains and canyons of the Wind River Range searching for some sign of gold. On several occasions they had to remain in hiding for days at a time due to the presence of Indians.

One evening in August of 1863, the three friends set up camp about fifty yards from a narrow, swiftly running stream, and quietly prepared dinner. They had learned to camp far from the noise of running water because the sound sometimes obscured other noises, such as those of approaching Indians.

Following the evening meal, Hurlburt carried the dishes to the stream to wash them. As he was finishing his chore, something in the stream bed caught his eye. Using a tin plate, Hurlburt scooped up a handful of gravel, washed it around in the plate, and extracted two large gold nuggets! Hurlburt called for his partners, and soon all three men were panning the gravel of the stream. They worked throughout the night under the light of torches, and by morning they had filled three pouches with nuggets of pure gold.

For the next several days, the men panned more ore than they ever dreamed they would find. Believing the gold came from a rich vein farther upstream, they determined to find it and retrieve even more gold than they were now taking in their placer operation.

About a month later, they discovered the vein. It proved to be very rich and the three decided that, with a few more months of digging the ore out of the rock matrix, they would be millionaires.

Because they had packed in an adequate supply of provisions, and because game was plentiful in the area, the three friends decided to remain in the canyon throughout the winter, working their placer operation and digging the gold from the rock. When the spring thaw arrived, they agreed, they would load their ore onto the pack horses and return to civilization.

To protect themselves against the harsh winters normally encountered in these high altitudes, the three men set to work constructing a log cabin on a flat spot not far from the stream.

The cabin was rude but adequate. Freitag constructed a large rock fireplace from the abundant flagstones found in the area. The hearth would provide heat and a place to cook. The roof of interlaced branches would protect them from all but the heaviest snowfalls.

Next to the fireplace they excavated a hole about two feet deep in which to store their gold. At the end of each day they would place the gold accumulated from the day's placer mining and excavation activities into the hole and cover it with a large stone. Hurlburt estimated the hole contained approximately a hundred thousand dollars' worth of gold nuggets.

One morning in the middle of October, Hurlburt remained in the cabin to chink the spaces between the logs with moss while Freitag and Smith left the cabin to work in the mine about a hundred yards upstream. The shaft had been excavated to a depth of six feet and the seam of gold grew thicker as they followed it. Around noon, Hurlburt heard rifle shots, but presumed one of the men had shot a deer.

When his partners failed to return to the cabin by sundown, Hurlburt grew concerned and went to look for them. On the way to the mine he called out their names but received no answer. When he reached the excavation, Hurlburt was horrified to find the mutilated body of his friend Smith—he had been scalped and all of his clothes save for his belt had been removed from his body.

Hurlburt looked around but could not find any sign of Freitag. Fearful the Indians might return at any moment, the terrified miner raced back to the cabin and gathered up his rifle and a few provisions. Just before leaving the cabin, he lifted the large stone from the hole, extracted two pouches of gold nuggets and put them in his shirt, replaced the stone, and fled.

For nearly a month Hurlburt wandered through the mountains, lost, cold, frightened, and hungry. He traveled at night and hid during the day. Because he couldn't see very well at night he had difficulty finding his way out of the mountains. He was afraid to shoot game for fear the Indians might hear him. Eventually he ran out of food and often went days without eating. Constantly traveling downhill in the hope of escaping the range, Hurlburt finally found himself walking along a road that led to Atlantic City, located near the southeastern tip of the range. A tinker driving a wagon found Hurlburt lying along the side of the road. Gaunt and delirious, his shoes worn to tatters, Hurlburt was suffering from frostbite on his fingers and toes. As the tinker placed Hurlburt in his wagon, he noticed that the semi-conscious man clutched tightly at two pouches of gold nuggets. By evening, the wagon had arrived in Atlantic City, where Hurlburt was treated for frostbite and malnutrition.

It took three weeks for Hurlburt to recover from his ordeal. When he did, he returned to Walla Walla where he lived with friends. Once spring arrived, however, he began to make plans to return to the Wind River Range, form a party of miners, and reenter the gold-rich canyon.

Because of the Indian menace in the mountains, no one would agree to accompany Hurlburt, so he foolishly entered the range alone. Remarkably, he found his way to the canyon and remained for three months living in the log cabin and panning gold from the stream. In September he rode into South Pass City with Smith's rifle and several pouches of gold nuggets stuffed in his saddlebags.

After showing the nuggets around town, Hurlburt once again attempted to organize a party of men to return with him to mine the gold, but as winter was setting in very few were interested. Discouraged, Hurlburt returned once more to Walla Walla.

The following spring found Hurlburt back at South Pass City. This time he successfully recruited four men and departed for the interior of the range. On this trip, however, Hurlburt was less fortunate. Time and again he became lost and could not find the canyon in which the rich, gold-filled stream and vein were located. Over and over he talked about the cabin, thinking he would find it in the next canyon, but eventually his four partners grew discouraged and the party returned to South Pass City after three months of fruitless searching. This time, Hurlburt announced he was giving up the search and returned to Walla Walla once and for all. Hurlburt's rich diggings were thereafter referred to as the Lost Cabin Mine.

In 1866, a man named Joe Poole wandered out of the Wind River Range with an intriguing tale. He had stopped in a tavern in South Pass City and heard several men talking about Hurlburt's Lost Cabin Mine in the Wind River Range. Poole became very interested and informed the men that he had just returned from a canyon that fit their description and had, in fact, found the skeleton of a man with a belt still tied around his mid-section near a narrow stream. Believing he had been close to the rich placer mine and ore vein discovered by Hurlburt, Poole bought a few provisions and returned to the mountains the next day. He was never seen again.

In 1886, a man named J.H. Osborne arrived at South Pass City after spending several months prospecting in the Wind River Range. He converted several large gold nuggets into cash, and when asked, told where he found them.

Osborne claimed he had been looking for color in the several streams he encountered in the range and finally found some in a shallow stream in a distant canyon. Spending several days in the vicinity, Osborne managed to accumulate several hundred dollars' worth of gold. He also found evidence of mining nearby, but neglected to examine the shaft closely. He told the listeners he stayed in an abandoned log cabin he found near the stream.

In spite of more than a hundred years of searching, no one has yet relocated the Lost Cabin Mine. The old log cabin has surely rotted and fallen down, but the sturdy rock fireplace has likely withstood the ravages of time. Several yards from the cabin runs a narrow stream replete with gold mixed in with the gravelly bottom, and farther upstream can be found the beginnings of an excavation which followed a rich, thick seam of gold into the rock. And adjacent to the old fireplace, in a shallow hole covered by a rock, lies a great fortune in gold nuggets, placed there more than a century ago by three men.

Lost Bighorn Placer

In the year 1865, a party of Swedish immigrants entered the Bighorn Mountains of north-central Wyoming in search of gold. Two years earlier the seven men had landed on the eastern shores of the continent, learned of the fortunes that hard-working and persevering men could make in the West, and decided to try their luck. They knew little about prospecting, but along the way they picked up knowledge and information as opportunities were presented and gradually learned a lot about ores and mining. When they entered the Bighorn Mountains, they brought with them two kinds of luck—good and bad. Good luck was with them when they discovered an incredibly rich placer mine in the mountain range; bad luck was inevitable, for the mountains were a refuge for the warring Sioux Indians and, unknown to the Swedes, had been declared off limits to all white men.

Riding across a sandbar in the middle of a slow-moving stream that bisected a broad meadow in the Bighorn Mountains, one of the Swedes noticed some sparkle coming from the ground. Dismounting, he inspected the sands and found them to be rich in flakes and nuggets of gold. Everywhere they looked on the sandbar, the Swedes found gold.

After nearly a year and a half of searching for the precious metal, the Swedes believed they had finally located the fortune they always hoped for. They quickly established a camp near the stream and spent every daylight hour panning for gold. Within the first week the men accumulated nearly twenty pounds of gold nuggets, flakes, and dust. At the end of each day they would divide the gold equally and place it in containers. As they sat around the campfire smoking their pipes, they spoke of the lives of luxury they intended to lead when they returned to Europe with their newfound riches.

The next day the snow began to fall and the Swedes decided it would be necessary to construct a cabin for protection against the freezing temperatures they knew would surely come. While two men continued to pan the gold, three commenced work on the cabin, while the remaining two went to hunt for game and lay in a supply of meat for the coming weeks.

Late that afternoon as the two hunters, each carrying a deer carcass, were returning to the cabin, they heard gunshots. Dropping the deer and crawling among the trees, the men approached the meadow and discovered their companions were under attack by Indians. As the defending Swedes fired from behind one wall of the unfinished cabin, the Indians rode back and forth in front of them, occasionally letting fly with an arrow. Knowing there was no way to go to the aid of their fellows, the two hunters remained hidden among the trees and watched.

Presently the Indians set fire to the cabin, forcing the Swedes to abandon its protection and run out into the open. One by one the miners were killed and their bodies mutilated. As several of the Indians scalped the dead miners, others ran into the burning cabin and retrieved some canned goods which they hacked open and ate.

For nearly three hours the two surviving Swedes watched the Indians from their hiding place, shivering from cold and terror. Eventually, the attackers rounded up the miners' horses and mules, mounted their own, and rode away to the west.

Cautiously, the two men crept toward the ruins of the smoldering cabin. Poking through the remains, they found their store of gold—several thousand dollars' worth packed in empty baking soda cans. Placing some of the gold in their packs, they quickly fled the site of the carnage toward the southeast, and several days later arrived at Fort Reno near the Powder River.

The Swedes related their plight to the post commander, who expressed little sympathy. He scolded them for ignoring the treaties which forbade white men to enter the Bighorn Mountains. The two Swedes, unaware of the existence of such treaties, pleaded with the officer to provide an escort for their return to the placer mine. When the officer refused, the Swedes attempted to organize a civilian party to reenter the mountains and mine the gold, but the army blocked their efforts and threatened to jail the two men if they persisted in their plans.

The Swedes remained at Fort Reno throughout the winter. When spring arrived they traveled to Rapid City, where they believed they would be able to recruit a group of men to return to the Bighorn Mountains. The Swedes explained to the men, about a dozen altogether, that though the mountains were off-limits to whites, the gold they would find in the sandbar located in the remote high meadow would make the risk worthwhile.

Two months later the party clandestinely entered the Bighorn Mountains and was never seen again. Unknown to the men, Sioux Indians in great numbers were at that time taking refuge in the Bighorn Mountains, and it is presumed that once they became aware of the presence of the miners, they killed them.

By the time the Indians were subdued and removed to reservations years later, most people had forgotten the tale of the rich, gold-laden sandbar in the Bighorn Mountains. Occasionally some old prospector would enter the range, hoping he would be the lucky one to stumble onto the site, but in the intervening decades the gold has never been found.

Ella Watson's Buried Fortune

For several years before she was hanged one summer afternoon in 1899, Ella Watson had buried approximately fifty thousand dollars in gold and silver near her residence in the Sweetwater River Valley in southwestern Wyoming. Watson's fortune, the proceeds from selling stolen cattle, has remained lost since that time.

Ella Watson was not the kind of woman normally found in frontier Wyoming. Experienced in robbery, rustling, and various hustles and cons, Watson was a large woman, a dead shot, and a match for any man. After running away from her family's Kansas farm when she was fifteen, Watson wandered throughout the West, eventually ending up in Wyoming. Moving from town to town, she developed a keen instinct for survival and quickly discovered how easy it was to separate a drunk man from his money. She eventually fell in with a gang of cattle thieves, and for several years pocketed an impressive amount of money from selling stolen livestock.

Watson's travels took her to the Sweetwater River country, where she met Jim Averill. Averill had arrived in Fremont County in 1885 looking for a job. Unable to find honest work, he started rustling cattle, driving them to another part of the state and selling them. During the next several months, he made enough money to build a saloon on a well-traveled road near the Sweetwater River. As his business prospered, Averill built a large corral behind the saloon in which to keep his stolen livestock. Small-time rustlers often stopped by Averill's establishment and, over a period of time, he got to know many of them and eventually hired several to steal cattle for him. The rustlers would take a few head of livestock at a time from local herds and when they had a sufficient number gathered, would drive them to Cheyenne to sell. All profits were returned to Averill, who would then pay off the rustlers.

Shortly after Watson met Averill, the two set up housekeeping in a cabin located about two miles from the saloon. Because more and more people traveled the Sweetwater River Road, and because the saloon became increasingly popular, Averill was afraid someone might recognize the stolen cattle he kept out back. He decided to construct a new corral, adjacent to the cabin, so the animals would be far from the curious eyes of his customers.

Averill was eventually appointed postmaster for this slowly growing area, and this additional job kept him busy to the point that he turned the entire cattle rustling operation over to Ella, and all transactions were conducted at the cabin. Ella would contract with the rustlers, receive the stolen cattle and place them in the corral, and when the time was ready she would hire some cowhands to drive them to the railhead at Cheyenne.

One afternoon, Ella looked up from her household chores and spotted a lone rider out near the corral. The stranger was slowly circling the large pen and staring intently at the livestock it contained. Presently he rode away and Ella went back to her work.

About an hour later, two men in a wagon followed by four more on horseback rode up to the front door of the cabin. As the dust settled around them, they called for Ella to come out onto the porch. When she did, they informed her that the cattle she had in her corral had been stolen from them and they had come to retrieve them.

After cursing the newcomers, Ella went back into the house and returned with a pistol. Just as she stepped out the door, two of the men grabbed her, wrestled her to the ground, and took her firearm away. Tying her securely, they threw her into the back of the wagon and rode to Averill's saloon.

With guns drawn, two of the cattlemen entered the building and returned with Averill. He, like Ella, was bound and then placed alongside her in the back of the wagon. Without locking the empty saloon, the entire party rode toward a cottonwood-lined bank of the Sweetwater River. Along the way, one of the men interrogated Ella about her cattle rustling activities and learned she had buried just over fifty thousand dollars' worth of gold and silver coins in a secret location near the cabin. The man said that if she turned the money over to them, she and Averill would be set free. Averill, listening in on the conversation, hissed at Ella to keep quiet and assured her the men were only bluffing and that no harm would come to them.

Several minutes later the wagon pulled up under a large, spreading cottonwood. As two of the men prepared nooses, Averill and Ella were pulled to a standing position in the back of the wagon. Ropes were thrown over a low-hanging limb and the nooses placed over the heads of the two prisoners. The leader of the cattlemen repeated the offer to spare their lives if they would reveal the location of the buried gold and silver. Ella was now very frightened and was about to tell him where to find the cache, when Averill once again insisted to her that nothing was going to happen and that the men were just trying to scare them.

After the nooses were secured around the necks of Averill and Watson, the men jumped from the wagon, leaving the two cattle thieves standing side by side at the ends of the ropes. When Ella and Averill declined for the third time to reveal the location of the buried cache, one of the cattlemen lashed at the wagon's team, causing it to bolt away. A few seconds later, Ella and Averill were swinging under the cottonwood limb, their lives being slowly choked from them by the constricting nooses. Within a few seconds they were both dead.

The cattlemen then returned to Ella's cabin and retrieved their livestock. While four of them drove the cattle toward the east, two remained to search the property for the cache, but could find nothing.

Over the years, others have traveled to the Averill-Watson cabin in search of the gold and silver, but all have reported failure.

About thirty years following the hanging of Averill and Watson, a letter from Ella to her family in Smith County, Kansas, was discovered in an old trunk. The letter, apparently written only a few weeks prior to her death, detailed her rustling activities and her love affair with Jim Averill. Obviously trying to impress the relatives she had fled years earlier, Ella, in a nearly illegible scrawl, told of the money she had earned by stealing cattle and how she had hidden a small fortune in an abandoned well near the cabin. Ella's letter, discovered in 1929, eventually came into the possession of a man who was familiar with the Sweetwater River country and the tales of local cattle rustling activities. He also knew the location of the old cabin in which Ella had lived.

When he arrived at the site of the Averill-Watson cabin, he was disappointed to discover it had been torn down and the surrounding land turned into pasture. Though he searched for hours, he could find no evidence of an old well. With the passage of time it had apparently been filled in and grown over with grass.

The gold and silver coins hidden a hundred years ago by Ella Watson still lie about six feet below the surface. Today the fifty thousand dollars in gold and silver coins that was buried during the 1890s would be worth considerably more. If someone were lucky enough to find the location of the old abandoned well, they would never have to work another day in their life.

Lost Ledge of Gold

Around 1870, Deadwood, South Dakota, was a bustling community catering to the miners who explored, prospected, and extracted gold ore from the nearby Black Hills. Deadwood was a collection of unpainted wood frame buildings and shabby tents in which prospectors and miners could find liquor, women, and gambling, as well as supplies. Many a prospector arrived in Deadwood bearing a sack of gold nuggets and announced a rich discovery someplace back in the remote and often dangerous hills. Many more wandered into the town discouraged and broke, having failed to find even a glimmer of gold.

One of the strangest tales of gold discovered and subsequently lost came from a grizzled old prospector named Boggs. The old man arrived in town one afternoon, disoriented and confused. He wandered up and down the main street as if trying to get his bearings, when he was recognized by another of his kind. Hailing the old man, his friend invited him into the nearest saloon for a drink.

At the bar, the companion asked Boggs how things were going out at his claim. The old man reached into his pockets and pulled out a handful of gold-filled quartz rocks and laid them on the bar. Several patrons quickly gathered around the old-timer to examine the gold; all were curious as to where it came from. When they asked him questions, the old man could only respond with a blank stare. Presently, he sat down at a table and related a most unusual story.

About a year earlier, Boggs had arrived at the Bear Lodge Mountains in northeastern Wyoming, just southwest of Devils' Tower, a prominent landmark. The old man panned the streams and inspected outcrops, ever searching for gold. One day, Boggs entered a narrow canyon he had never seen before and worked his way up toward the head of it. As the canyon narrowed and the incline increased, Boggs, tired from the climb, found a fallen log on which to rest. Suddenly he noticed a thin ledge of different-colored rock protruding from one wall of the canyon. On closer inspection, he happily discovered the ledge was an outcrop of gold-laced quartz. Boggs dug out several small pieces and found them to be the purest gold he had ever seen. The vein, according to the old man, extended for several feet in either direction. Truly, he thought, here was enough gold to make him as wealthy as a king.

For several months Boggs dug the gold from the quartz matrix, accumulating an impressive amount of the ore. When he was not working his mine, he busied himself with the construction of a small, crude rock cabin at the mouth of the canyon. Behind the cabin, Boggs dug a shallow trench in which he placed and covered up his gold.

Deadwood was located some twenty-five miles to the east, and it was during a trip to town to purchase some supplies that Boggs met a young lawyer named Burns. The old man and the lawyer became friends and Boggs would always stop to visit during subsequent journeys into town.

During one visit, Boggs asked Burns if he would like to accompany him into the mountains. At first the lawyer demurred, but Boggs was insistent and finally the young man gave in. The next morning, the two rode out of Deadwood and into the Bear Lodge Mountains, where Boggs told the lawyer about his gold discovery and took him to the ledge. Burns knew very little about mining, but he recognized gold when he saw it. Staring at the ledge, Burns saw plenty of the ore and suddenly realized Boggs was a very rich man.

After remaining with Boggs for nearly a week, the lawyer returned to town, his pockets filled with several pieces of gold-bearing quartz from the old prospector.

A few weeks later, Boggs decided to take a break from mining and go deer hunting. He had seen plenty of deer sign in a neighboring canyon, so he decided to ride over there. While riding along a particularly rocky stretch of ground, Boggs's horse slipped on some loose stone, panicked, and threw his rider. Boggs landed on his head and was knocked unconscious for several hours. When he finally awoke, it was dark. Looking around, Boggs had no idea where he was. Leaving his gun and hat on the ground, the old man rose and walked out of the canyon. Daylight found him wandering in the range, searching for some familiar landmark, but the blow he received on his head apparently had affected his memory. He was unable to find his way back to the cabin and the canyon that contained his rich ledge of gold and his buried fortune. Days later Boggs—tired, hungry, and filthy—somehow wandered into Deadwood, where he was recognized.

One of the men at the saloon knew that Boggs was a friend of Burns's and sent for the lawyer. When Burns arrived, he corroborated Boggs's story of the rich gold mine. Taking Boggs home with him, he placed the old man under the care of a physician and saw to his comfort.

As Boggs recovered during the next few weeks, he continually expressed a keen desire to return to the Bear Lodge Mountains and relocate his gold mine. When he was well enough to travel, Burns outfitted the old man with a good horse, a pack mule, and adequate supplies. Three weeks later, however, Boggs was back in Deadwood. He informed Burns that he could remember nothing about the mountains or the location of the canyon, that he had become lost and confused, and had finally given up.

Burns decided to accompany Boggs on a return trip to the mountains in search of the gold, but like the old prospector, he too became lost and confused. After a month of fruitless searching, the two abandoned the mountains and returned to Deadwood.

Over the next few years, several others made attempts to locate Boggs's mysterious canyon. At one time, there were nearly a dozen prospectors combing the Bear Lodge Mountains searching for a small rock cabin at the entrance to a narrow canyon.

One man did find the cabin, but at the time he was unfamiliar with the tale of Boggs's lost gold ledge and hurried caching of the ore. As he explained it in later years, the prospector had spent several weeks in the Bear Lodge Mountains panning some small streams. Quite by accident, he entered a narrow canyon he described as little more than a ravine. Just at the entrance of the canyon, he said, he discovered a low, crudely constructed rock cabin. So well did the structure blend into the surroundings that the prospector was within ten feet of it before he recognized it for what it was. Thinking someone might be about, he called out several times but received no answer. Assuming the canyon was the domain of whomever lived in the cabin, he rode away.

In the annals of Western history, the Black Hills of South Dakota were judged as one of the richest gold-bearing locations ever discovered on the North American continent. The Bear Lodge Mountains, located just across the border in Wyoming, are part of the same igneous core and share the same characteristics as the Black Hills. Though prospecting and mining in the Bear Lodge Mountains never compared to that which took place in the Black Hills, the gold ore that was occasionally discovered there proved to be quite pure and assayed at top dollar. Old man Boggs undoubtedly stumbled onto an incredibly rich lode in the Bear Lodge Mountains, but as a result of bad luck, lost it.

To this day Boggs's lost ledge of gold has not been found, nor has the cache of ore he buried behind his rock cabin.

Section 2:

TREASURE TROVE LAW

This section is short and to the point. Be very careful and do not talk to anyone about finds until you totally check out all laws.

The Right to Search on Private Land

To search on private property, obviously the first thing the treasure hunter must do is get permission from the owner of the property. Although this permission is often given verbally, any seasoned treasure hunter, whether he's seeking a jackpot treasure or simply bottles and coins, prefers a written agreement with the property owner. In general, such an agreement should include the following:

1) A statement about the reason for digging.
2) A statement concerning the ownership and/or division of any treasure found.
3) A guarantee that the searcher will leave the property in the condition in which he found it. (In other words, he'll fill in the holes he digs.)
4) A disclaimer of liability on the part of the property owner if the treasure hunter meets with an accident while searching for the treasure.

If you write out your own agreement, as many treasure hunters do, you should make at least one carbon copy. Searcher and property owner should each sign both copies.

Ready-made forms are available at various metal detector and treasure hunter supply stores. Many clubs provide them for their members.

The Right to Search on Public Lands

Some areas of public land, federal and state, are absolutely barred to treasure hunters. National Parks and most State Parks are in this category. National Forest, in general, are open to treasure hunters, but new regulations are being formulated in regard to some National Forest lands. You should check with the office of the particular forest in which you intend to search, or with one of the regional offices listed elsewhere in this Information Center.

The Bureau of Land Management, which owns millions of acres of land (the amount changes, due to sale or transfer of some of it to other governmental units), in general has no restrictions on treasure hunting on BLM land.

Sunken Treasure

Inasmuch as most of our readers will not be going after this type of treasure, we will not attempt to deal extensively with it here. The laws pertaining to salvage are complex, and the rules in regard to ownership or rights to the valuables found aboard sunken ships are too difficult to summarize briefly. We recommend Salvage Law from Sunken Ships to Outboards, a handy little manual that will give you most of the basic facts.

Who Owns Treasure You May Find

If you find a treasure trove, can you keep it? This question has had varied answers in the courts, which have heard many disputes. Some cases in the U.S. have resulted in legal definitions that tend to set up the rule of "finders keepers."

Today your right to keep treasure you find, or to carry out any agreed-upon split with the owner of the property on which you find it, is modified by state laws. Each year more states are passing laws which govern the disposition of treasure. Such laws are generally referred to as "Antiquities Acts." Louisiana's Act 172 is an example. It reads, in part:

Florida has a law entitling it to one-fourth of the proceeds from any treasure found there. It is necessary in this state, as in some others, to obtain a license.

Regulation of treasure hunting and its proceeds is an area which is subject to an increasing number of laws, so the situation in any given state may change. An excellent general source of information on recent laws is A.T. Evans's Treasure Hunter's Yearbook. It is published early in each calendar year and contains a section on "Law and the Treasure Hunter." A copy may be ordered directly from the publisher, Eureka Press, Odessa, Texas 79760. The price is $4.00 postpaid.

Specific local information pertaining to your state or locality can be obtained at local metal detector dealers, rock shops, and treasure hunter and rockhound clubs.

Taxes

Two questions often asked by treasure hunters:
"Can I deduct expenses connected with treasure hunting?"
"What taxes do I have to pay on treasure I find?"

We won't attempt to go into the whole complex matter of income taxes. The IRS has changed the rules, and its interpretations of the rules, many times. So before you actually file a tax return, you had better check with the IRS for its latest thinking on this matter.

We can present you with a few general rules, however. The basic one pertaining to deductions for expenses connected with a hobby--any hobby, not just treasure hunting--is that you are entitled to deduct expenses equivalent to the income you derive from it. In other words, if you expenses connected with treasure hunting were $800 in a given year and in that year you found treasure worth $500, you could list $500 as expenses, and you wouldn't pay taxes on that amount.

If you're conducting your treasure hunting as a business, with the full intention of earning money from it, you can deduct all legitimate expenses--even gasoline needed to get you to treasure sites, lodgings while you hunt for the treasure, etc. Metal detectors, gold extracting equipment, scuba gear, books, and possibly even a four-wheel-drive vehicle are deductible expenses.

As to taxes you pay on treasure you find that how much they are and when you pay them depends on the kind of treasure. If what you find is gold (other than nuggets) or currency, the IRS quite naturally views it as income and takes a dim view of anyone who doesn't count is as such. However, if your treasure is in the form of bottles, barbed wire, or other collectors' items, a different situation pertains. Suppose you find a bottle that is worth $200, but you don't sell it. It's not, in the ordinary course of things, counted as income. It becomes income only if and when you do sell it.

Whatever its nature, if your find should be a big one, you can simply keep it and dole it out to yourself, counting as income only that portion of it turned into cash in any given year. Even if your find is in the form of currency, it is sometimes possible to arrange to spread out the income from it over a period of five years, under the IRS rules for "income averaging."

But we say again: Don't take our word for any tax information. Consult with the IRS or your own tax expert.

The ancient Spanish inscription says "You are standing in quicksand, stupid".

Section 3:

READING SPANISH SYMBOLS

This is a good basic introduction on how to read Spanish treasure symbols. As always, there are many variations there of. On the trees that are still alive with these marks, they are very old and hard to see, so look hard. I've even painted some with white non-toxic paint so I can read them. When I was finished, I washed them off. Most symbols are not complete without the others that were meant to go with it. Usually you look off to the right of each symbol to get the next one. The Spanish were very smart and accurate. You need a good compass. Remember that true north has shifted a little since the 1700's.

This section on Spanish Treasure marks and mining symbols came from many sources, most of which I don't have information on who first printed them. Some came from a young man by the name of Stephen Shaffer. I spoke with him on the phone and got his permission to use some of his information. Thank you, Steve.

Two more 'Flower' symbols. These two symbols may look like each other at first glance but they are really two completely different symbols. One is showing a canyon, the ridges and just below and to the left of the top of the ridge is a mine. The petals indicate how many 'Varas' to proceed after reaching the head of the canyon, here the mine can be found. The other symbol is showing a canyon with two side knolls, and a large mountain at the head. The mine will be near one of the knolls. Notice how the flowers differ. When searching for symbols the flower symbol is most important if it is accompanied by a 'Notched' tree. If there is no notched tree then proceed ahead until one is found or another symbol is found.

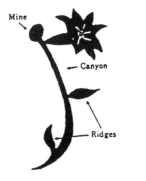
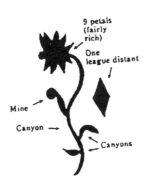

120 Degrees

120 Degrees

120 Degree

Travel on, to a
triangle marked out
by trees or rocks.
Treasure will be at
one of the corners.

#1.Travel to the right
around bend, or sweep
around, away from the
triangle.
#2.Treasure will be
buried where 'Arc'
connects to triangle

The longer arm means
to travel around 120
degrees. The rest of
the symbols reads
the same as the
symbol to the left.

Change direction; Go
to your left around
the nearest hill. If
reversed, the trail
goes to the left.

Kings mine

Turn about or route to

One league

Mine

Mine with one entrance, and treasure

Mine or cave

One league

One and one-half league

Living quarters

Wrong way, change direction, Return to last marker

Ladder: Mine or mine shaft.

Treasure in tunnel.

Mine shaft, tunnel or cave.

(1) In a tunnel.
(2) A tunnel with only one entrance.

Mine shaft, tunnel or cave.

(1) In a tunnel.
(2) A tunnel with two entrances.

Tunnel or shaft.

(1) Mine shaft.
(2) Water well.

(1) Mine shaft—filled or covered.
(2) Water well—filled or covered.

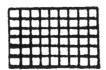

(1) Concealed opening.
(2) Concealed mine or cave.

(1) Three varas deep.
(2) Proceed three varas.
(3) Treasure divided into three parts, in boxes or chests.

Shaft or cave—filled or concealed.

(1) Shaft or tunnel at foot of hill and center of hill.
(2) Shaft or tunnel in center of draw, meadow or valley.

Sombrero:
(1) Number of sombreros indicate number of Spaniards or Mexicans in expedition or sometimes the number killed.
(2) Crown of summit; the top.
(3) Under rock.

(1) Treasure under.
(2) Mountain crest—dot indicating location of mine or treasure.

Stream to be crossed at this point, then continue in same direction as before.

Stream to be crossed at this point, then continue in direction indicated by the extended line (right).

Treasure is buried at the junction of two streams farther on.

Treasure is buried in the middle of a stream; continue in same direction.

Treasure is in a box or chest and is buried in the center of *this* stream.

Treasure is buried in the middle of *this* stream. This is the treasure site.

Go ahead; treasure will be found in a stream with steep banks on both sides. (If two wavy lines and a straight bar; stream with one steep bank—the steep bank indicated by the straight bar.)

This is the treasure site. (Location described at left on another marking.) Treasure is buried on the bank in the angle indicated by the dot or circle.

Treasure is buried in or near stream which is dry certain times of the year.

Proceed to nearby stream where you will find a waterfall. Cross at this point and continue in same direction.

Distance Indicator Symbols

Varas: Two.

Trail: Correct line, continue.

(1) One vara.
(2) Negative; minus.
(3) Horizontal; even.
(4) Vinculum.
(5) 50 or 100 varas.
(6) Creek or canyon.

(1) Double the distance.
(2) Two varas (Four varas, doubled)
(3) Equals.

Travel around bend, away from second
hill. Treasure buried where arc ends.

Treasure buried in
lower half of hill.

Treasure buried between two small
hills or knolls.

Gourd or Key; Follow the long arm.

Slight change in course, to the left in this case.

Dagger; Points to treasure or mine. If the blade is separated from the handle, it means to return to the last symbol and go in the opposite direction.

Cannon; Follow the Barrel of the Cannon, or go up the nearest canyon or hill.

Direction Indicator; Proceed straight ahead. If the symbol is tilted either left or right, follow the tip of the symbol inlake. the direction it is pointing.

Fish; Direction Indicator. Trail is next to or in a creek or on the edge of a lake.

Peace Pipe; Direction Indicator. Follow in the direction of the

On trail to wealth; Follow the long arm of the symbol.

Direction Indicator; Follow
the barrel of the hand gun.
Usually found near a mine.

A simple form of follow
marker, usually pointing
to water.

Direction Indicator; Follow the
long arm but swing to the left.

This symbol is not
only a direction
indicator, but a
symbol that means that
a mine is close by.
Also can mean mineral.

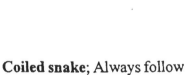

Coiled snake; Always follow
the head.

Turtle: This symbol is a famous symbol often used by the Spanish. It always means treasure nearby. If the head is removed from the body this means the treasure is gone. If the legs are removed from the body this means you are at the site. If the tail is not pointing straight down or back from the body but is curved away from the body then follow the direction of the tail. Also look at the back of the turtle because sometimes there may be symbols on it as well.

Turtle with no legs;
Go no further, you
are at the treasure.

Head points toward treasure.
Continue straight ahead.

Feet close to the body;
Go ahead slowly, near
treasure.

Turtle with diamond back;
Count diamonds, if four, go
four varas or four leagues.

Turtle with head removed;
Treasure is gone, removed.

Turtle with tail curved;
Go in the direction of the
tail.

This symbol is the meaning for 'Forest' or it's telling you to go into the trees for more symbols.

This symbol is a Mexican cross, mostly used as a follow marker. To follow This symbol, look slightly to the right and then proceed ahead.

Follow Markers

The Latin Cross; Mostly used during the early expeditions of the 16th, 17th and 18th centuries. However, it was used somewhat later on.

Spanish Cross; Used more in the 19th and 20th centuries.

Ninus; Always follow the point.

Common Directional; Follow the tip of the letter.

I thought of how Jim Bridger once told a prospector that there was a diamond on top of a mountain in the Yellowstone country which could be sighted fifty miles away "if a man got the right range on it when the sun was right." The prospector offered the old scout a fine horse and a new rifle if he would show it to him.

Notwithstanding the advantages of shadows, buriers of treasure have from time immemorial generally marked their deposits with signs that are directly visible, though usually quite enigmatical. The Spaniards, since they buried most of the treasure in this country, developed, of course, the most elaborate code of symbols. A register of these signs, collected and deciphered for the most part by a lawyer who refuses to allow the disclosure of his name, is now—for the first time, I believe—printed.

X   *On line to the treasure. X is also a common designation on landmarks.*

✝   *Cross and other rich objects pertaining to the Church are buried here.*

The cross might mean many things. So potent was its symbolism that the very sign of it might protect a man's possessions as well as himself. It often said: "I have been here"; "A Christian has passed this way." When Coronado went east in search of the Gran Quivira, he gave instructions that he was to be trailed by means of wooden crosses which he would erect from time to time along his route. Again when Fray Marcos de Nizza set out in 1539 to hunt the Seven Cities of Cíbola, he had instructions, should he find himself on the coast of El Mar de Sur (the Pacific Ocean), "to bury written reports at the foot of a tree distinguished by its size, and to cut a cross in the bark of the tree, so that in case a

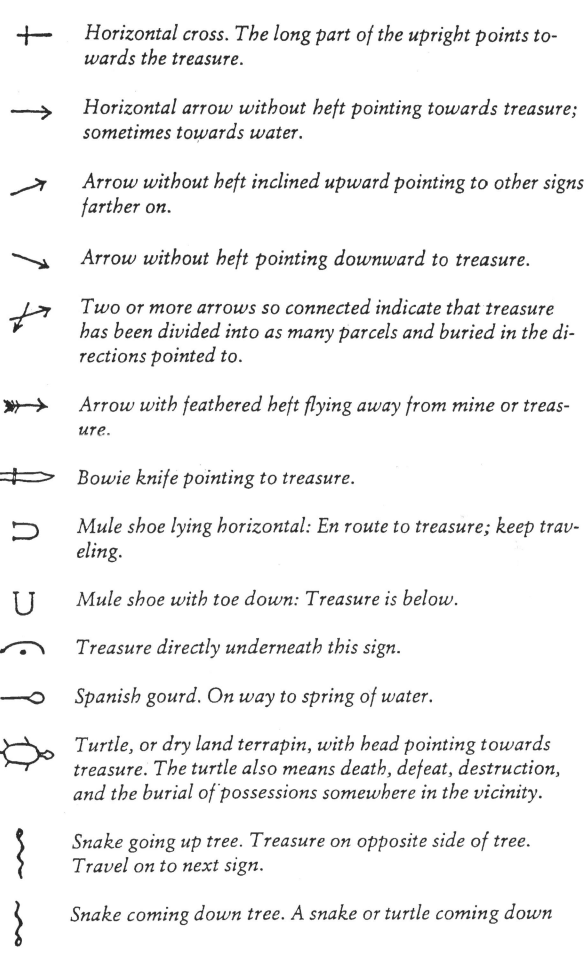

ship was sent along the coast, its crew might know how to identify it by that mark."

Horizontal cross. The long part of the upright points towards the treasure.

Horizontal arrow without heft pointing towards treasure; sometimes towards water.

Arrow without heft inclined upward pointing to other signs farther on.

Arrow without heft pointing downward to treasure.

Two or more arrows so connected indicate that treasure has been divided into as many parcels and buried in the directions pointed to.

Arrow with feathered heft flying away from mine or treasure.

Bowie knife pointing to treasure.

Mule shoe lying horizontal: En route to treasure; keep traveling.

Mule shoe with toe down: Treasure is below.

Treasure directly underneath this sign.

Spanish gourd. On way to spring of water.

Turtle, or dry land terrapin, with head pointing towards treasure. The turtle also means death, defeat, destruction, and the burial of possessions somewhere in the vicinity.

Snake going up tree. Treasure on opposite side of tree. Travel on to next sign.

Snake coming down tree. A snake or turtle coming down

a tree means that treasure is on that side of it. Measure distance from the tip of the reptile's tail to the ground. Step off ten times that distance straight out. At the termination of the distance stepped, one should find either the treasure or another sign.

Snake in striking position with head pointed toward treasure.

Snake coiled on tree or rock indicates presence of treasure directly beneath.

A straight line indicates a certain number of varas to be measured off, the vara being 33⅓ inches; the number of varas called for usually ranges between 50 and 100.

Two straight lines indicate double the distance of one line.

Flight of steps. This sign indicates that the treasure is down in a cave or shaft.

Treasure is to be found within a triangle formed by trees or rocks.

Over deposit, which is located within a triangle made by trees or rocks.

Triangle formed by trees or rocks enclosing treasure.

Triangle formed by trees or rocks with treasure in the middle.

This sign indicates that while the deposit is marked by a triangle of trees or rocks, it is to be found to one side of the triangle.

Deposit is around a bend or curve away from triangle formed by trees or rocks.

Treasure buried in box or chest.

Shadows and Symbols

Peace pipe. Friendly Indians.

Sombrero, or hat. The number of sombreros shown indicates how many people were in the party that buried the treasure. The sombreros may also indicate the number of men killed by an enemy.

Mines close by. Any representation of the sun indicates proximity of mineral wealth.

ORO   *Oro (gold) is short distance away.*

G   *Gold short distance away.*

[*A tunnel.*

Stop; change direction.

or   *Perhaps variant signs of the cross.*

Greyhound. As to the meaning, there is some doubt.

Section 4:

PLACER GOLD LOCATING

Just a few nice photos showing the best places to look for gold in streams.

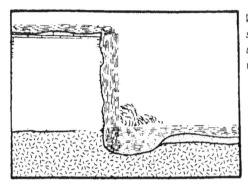

Waterfalls with considerable turbulence usually pulverize and wash out gold.

Waterfalls which have stream bed sand beneath them almost always concentrate gold.

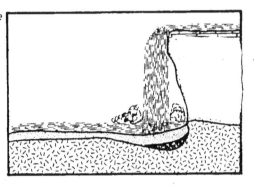

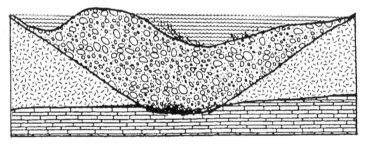

Sand bars often have a recent accumulation along the sides, but to get any quantity of gold, the best solution is to go to bedrock.

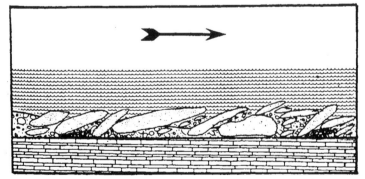

Rocks "shingle" on a stream bed near bedrock. An accumulation can be a good gold trap and many have given up a considerable amount of placer gold.

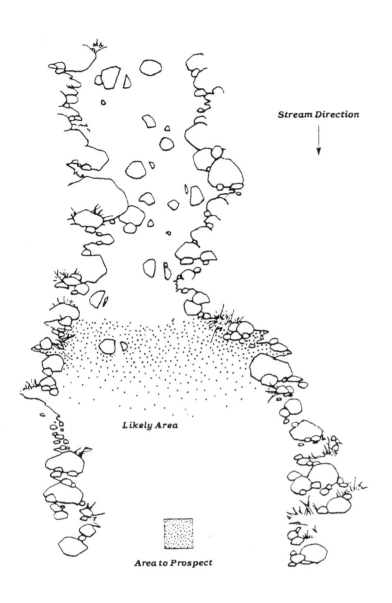

Stream Direction

Likely Area

Area to Prospect

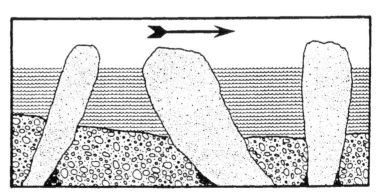

Boulders extending some distance into the stream bed are always worth investigating. Gold can accumulate downstream, upstream or on both sides, so take nothing for granted.

PLACER PAYDIRT

Points to remember: Look first to the obvious: large cracks, crevices, fissures (particularly those at points where the river or stream will slacken, or wide points, changes in direction, etc.), and the inside bends in the curves of the stream (more than likely associated with deposits of gravel). Observe areas where the river slows down for any reason (where the course of the river widens, after a set of rapids, deep pools in the riverbed or where the gradient dissipates thereby slowing flow). Check downstream from intersections of fault lines. Notice sections of river wherever the water current might slacken because of obstructions (boulders, outcroppings of bedrock, etc.). Be on the lookout for concentrations of black sand, old nails, horseshoes, etc. Sample depressions in rocks, particularly those that have many small boulders jammed into the depressions. Check and take a bucket of the material that is caught up in the tangle of exposed tree roots at the water's edge. Make notes of areas in the midsection of the stream where quantities of large boulders have drooped at the point at which the river or stream has widened. Pay attention to any area or condition that would have caused the carrying power of the stream to be reduced. Also look for large boulders that, although "beached" now, were at some time surrounded or completely submerged by a long-forgotten spring flood. Winter or early spring survey trips will reveal mid-stream boulders high and dry and easily accessible by mid-summer. It also pays to check out areas where there's evidence of placer tailings from the "old" days, where miners in the desire to cover as much ground as possible often exhibited "haste made waste" in their recovery methods.

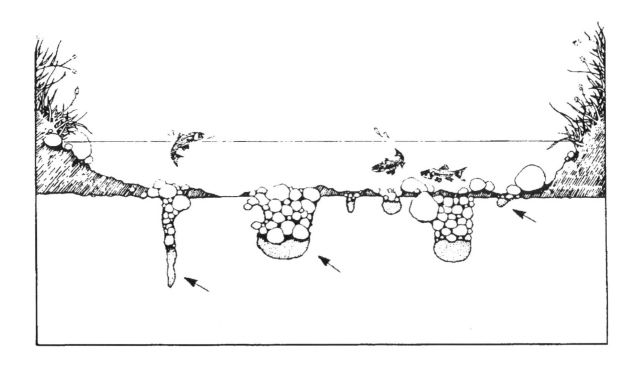

O.K., that's all the treasure- now drop the rope down to me, Burt.....Burt....Burt!

Section 5:

DOWSING

This subject is very interesting. It is a natural skill that can be developed by almost anyone. Use common sense and stay balanced in all your activities. I have seen many dozens of books on dowsing. Most were very confusing. The only one I ever saw that I could tune in to was by Ona Evers. I used some of her statements in this section because she worded them so perfectly. I could not change them, or think of a better way to say it. I would love to talk to her some day. I have been unable to find her.

Just what is dowsing, anyway? Basically, dowsing is searching. Searching for, among other things, under ground water (so-called water witching), minerals, oil, lost objects, people, information.

It is used quite matter-of-factly all over the world to locate underground cables, gas pipes, and water lines. Archeologists use dowsing to find hidden sites. Building industries use it to pin-point structural weaknesses. New uses for dowsing are constantly being discovered, and old uses revived. Although often surprising, dowsing is not weird or spooky. It is a useful art or skill that anyone can learn with practice.

Some persons grasp the art instantly- so fast that it's as if dowsing grasped them. They are the truly gifted ones. The rest of us have to try harder. Dowsing's as natural as memory.

In Europe, particularly in England and the Soviet Union, dowsing is used as a commonplace adjunct to archeology. Roman camps, druid ruins(in England), demolished palaces and old battle grounds (in Russia) are being reconstructed for modern eyes with information found by dowsing.

Meanwhile, millions of producing wells over the centuries have been dug, minerals and oil found through dowsing. So avoid the arguments. You'll never convince a true skeptic, no matter what.

If you follow the simple suggestions in these pages, and the L rods swing out like garden gates, then you are a strongly gifted dowser. I envy you, but not so much that it hurts. The rest of us, the weaker gifted, will be following right along.

Why should you even bother to learn to dowse it your gift is weak or almost zero? Because you want to . And because it's intriguing and fun, and useful. When you've developed the knack, you'll find it's as satisfying as an athletic endeavor or creative effort that foes just right. Dowsing can become the most absorbing hobby you've ever learned, or even a profitable vocation.

If you try the methods given in these pages, practice, and keep an open-mended venturesome spirit there is nothing to prevent your becoming a true dowser in a short time.

The Rods and Their Uses

Gold, silver, ivory, horn, rare woods, crowbars, snips of wire, fishing rods, keys, welding rods, nails, copper, nylon, maple, apple, peach, hazel, willow woods, whale bone, buttons, car radio antennae, blades of grass, plumb bobs, coat hangers, pendants and pliers- these are some of the materials and objects used in dowsing rods.

All of which means that the rod doesn't matter. It is you- the dowser- that matters. Many dowers use no rod at all. Just their hands. The rod is merely a pointer, an indicator like the needle of a compass. Something you can see. Any indicator which can be held in "unstable equilibrium" (13) will work. (I'm assuming that you're not quite in the Hand Dowser class yet.)

When describing them, most dowsers divide the fords into four main classifications:

1. The L rods, which are any rods that swing outward or inward as a recognizable signal - like the coat hangers.

2. The pendulum - which can be any object at all hung from a cord, or such like.

3. The bobber - a straight flexible stick, thicker at one end than on the other. When held by the thin end, its up-and-down bobbings are used for information - often something that is to be counted, or measured, such as distance or depth.

4. Last is the Y- shaped rod, the classical symbol of dowsing. It , too, can be made of any material. However, a flexible rod is felt to be more sensitive and easy to "read", although several the local dowsers use stiff and well-aged branches.

Any of the four rod types can be used in place of any of the others, depending on your preference.

The best way to get started is to get two L rods made of cast hangers or bent welding rods. Hold them like two six-guns. Hold them level and walk smoothly. Beforehand have someone hide something in your yard. Or you can look for your own water pipes. Practice on something you know is there. You might have to practice for a whole month before the rods swing out or in for you reliable . If you are gifted it will happen soon and with bigger movements. When the rods don't move, there is usually something blocking their action. In the case of beginners, it is almost always simply a lack of concentration. We're letting thoughts from the conscious mind force their way in. Be patient. The knack will come.

Have you ever seen an artist squint up his eyes to blur things and fuzz out obtrusive details from the overall pattern of the of the scene before him? I've found this squinting is helpful in keeping my mental focus on the object of my search. However, if you feel you've lost the spirit of detached anticipation, stop. Dowsing is a skill that, for me at any rate, is best practiced in small frequent doses.

Many dowsers say it doesn't work for them unless there is a purpose to it, a certain importance in the hunt. Something they, or someone else, really needs. Obviously, there isn't much importance in finding what ever your friend hid for you in your yard. What is important is that you are searching for a new skill.

After you practice while and start to get fairly good you might be tempted to go show off for someone or take part in a test for science of t.v. never do it . For some reason it never works and you will be portrayed as an idiot.

Lets get back to a little practicing. Remember think of what you're looking for. Concentrate visually on the tip of whatever rod you're using, and either picture the object of your search, or ask, "Am I over a water line?" Or more specifically, "please tell me when I'm over a pipe with running water in it." Be polite to your rod. You will get exactly what you ask for .

Relax, focus only on the tip of the rod, expect nothing, but try to be athletically poised. Slowly the rod will swing around to point at the hidden cruet. And then, again, it may not. Don't be discouraged. The knack will come. Remember when practicing search for anything you can verify.

Don't forget you might get to what is called "learners Plateau". All of a sudden you don't get any reading at all. It happened to me. As I started to get the gang of it, I had become careless and hadn't really focused on the object of my search. Instead, I'd lit myself become tense all over, trying too hard. Every athlete, or creative person, knows that's the dumbest thing you can do. Teeth clenched, muscles tight, energies going in the wrong direction.

So unclench and think only of what you're looking for. And don't knot your brow about that either. Flow with it. Enjoy it. You don't have to dowse. No world shattering events will take place it you don't. No one is waiting, shovel in hand, for you to point out the spot for him to dig. I hope.

Remember that what ever makes dowsing work can be effected greatly by other people. Make sure you are not surrounded by negative people or noise and that you are not to tired. Always try to avoid all the fancy gadgets out there that are supposed to help you dowse better.

Keep your dowsing as simple and uncluttered as possible. Passive concentration is the watchword. Don't push against the river. Float.

PRINTED WORKS USED

Legends of Lost Missions and Mines - C. W. Polzer S. J.

Early Spanish Treasure Signs - W. Mahan

T. H. Northwest - Ruby Hult

Handbook of Treasure Signs - M. L. Carlson

Prospecting, Placer - Joe Petralia

Everybody's Dowser Book - Ona C. Evers

Treasure Monuments. Archives Assisted - C. A. Kenworthy

Treasure Hunters Manual #7 - Karl Von Muller

Principles and Practice of Radiesthesia - Abb Mermet

Reading Spanish Symbols - Stephen Shaffer

The next two chapters are related to the previous chapter on treasure marks. The next chapter is on mine trail markers/monuments. Read carefully. It may help you greatly next time you are out looking for treasure. This information is being used by permission from Chuck Kenworthy. I called him back in 1997 and talked with him. He was very nice. He has books out that go into even more detail on these subjects. The second chapter is on Spanish Death Traps. This is very important to know.

Section 6:

SPANISH TRAIL MONUMENTS
Introduction

Beginning in early 1983, we began receiving hundreds of drawings from archives that depicted the trail markers and monuments that Spain required to be built/constructed both into and out of mountain/hill ranges that contained a major mine or treasure.

I would venture to say that tens of thousands of Spanish treasure/mine trail monuments/ markers exist today in the United States. I have seen well over one thousand of them throughout our country as well as hundreds just over the border in Mexico. The King of Spain **ORDERED** all treasure/mine trails to be monumented according to Spain's drawings of markers/monuments. The Palace of Governors in Mexico City and Sante Fe (New Mexico) **INSTRUCTED** the exploration groups, haciendas, mining/explorers etc. in the basics of both monument meanings and map codes. **AFTER** a mine was established, the Palace of Governors would appoint two marker/monument building supervisors to oversee the actual construction. Additionally, the Palace appointed a map-maker and a "religious" from the nearby or adjoining cathedral to begin accompanying the hacienda/miners for the purpose of caring for the souls of the miners to, from and at the mine. Also, of course, these appointed "religious" had freedom to map Indian villages and convert natives they might encounter. Note: The Jesuits were firmly against Spain's use of natives as laborers because of Spain's extremely harsh treatment of native labor. The Jesuits expulsion from New Spain in 1767 was primarily caused by their opposition to Spain in this matter. Both Mexico City and Santa Fe regulated and enforced the King's rule, received the King's 15% to 20% fee from the haciendas/miners etc., required the trail monuments to be built so that if they wanted to "inspect/check" the operations without notice, they could easily find and follow the monumented trail into desolate mountainous terrain. Also, if all miners were to meet with some great disaster, the King of Spain could again located and re-establish the mine - or retrieve the hidden treasure by following the monumented trail.

Therefore, the trail markers to and from were required as well as treasure/mine maps. Note: All treasure/mine maps symbols/signs etc. were also identical in use and meaning throughout this New World. Also, a standard or special list of "measurements" were used on treasure/mine maps because Spain could not operate with hundreds of different codes, measurement and different monument/marker meanings when they were dealing with so many mines in the New World.

If we think about it for a minute, it becomes very clear that Spain was extremely wise

Foreword

This is a photobook of carved/cut Spanish markers and monuments on **TREASURE TRAILS** existing in these United States. Also their **MEANINGS** and how to **VERIFY** that they are not the work or design of **MOTHER NATURE** will be shown

Even though some of these **TREASURE TRAIL MARKERS/MONUMENTS** have been around for 300 years or so, they have not been recognized and are being shown and explained herein for the first time through both archival and **in-the-field RESEARCH**.

RECOGNIZING and **UNDERSTANDING** these trail **MARKERS** should be of significant interest to, **HISTORIANS, ARCHAEOLOGISTS, HIKERS, BACKPACKERS, COWBOYS, ROCKHOUNDS, HUNTERS OF LOST MINES AND TREASURES**, as well as any that might roam the deserts and mountains while **FISHING, HUNTING** or just **VACATIONING.**

Some of these Spanish monuments come close to rivaling those of **EGYPT** and **PERU** in their detail. Properly understanding the **MEANING** and of course, **RECOGNIZING** these **TREASURE TRAIL MONUMENTS**, could lead anyone, even someone driving a car, on his way toward one of the many **HIDDEN MINES OR TREASURES AS YET UNFOUND IN THESE UNITED STATES**

TREASURE TRAIL MARKERS and **MONUMENTS** were constructed according to Spain's official directions and instructions here in the new world. **THAT IS WHY THE MONUMENTS FOUND TODAY IN KANSAS AND KENTUCKY ARE INDENTICAL TO THE ONES EXISTING IN CALIFORNIA, ARIZONA, TEXAS, COLORADO, UTAH, NEW MEXICO, MEXICO, ETC., ETC., ETC.**

Ladder: Mine or mine shaft.

Treasure in tunnel.

Mine shaft, tunnel or cave.

Mine shaft, tunnel or cave.

(1) In a tunnel.
(2) A tunnel with only one entrance.

(1) In a tunnel.
(2) A tunnel with two entrances.

Tunnel or shaft.

On the opposite page are two photos of the same cross. Most trail crosses are simply carved/cut into cliff faces and large boulders along the **OUT** (homeward) **TRAIL**.

NOTE: Be very alert to the fact that a **CROSS** or a **"T"** (a **TOBIAS SYMBOL**) on a map means exactly the opposite of what a cross means when found in the field along trails.

#1 is a cross made by enlarging a cave or crack or actually digging it, which they probably did because of it's major messages and it's all important location. The **CROSS** itself just says that we are on the homeward trail.

#2 is a shaped 200 pound rock that serves two purposes. **"A"** This rock is wedged and cut to fit into the lower part of the cross and has a **POINT** that points down the curved bottom line of the cross saying **FOLLOW DOWN**. We follow down and the very bottom of the cross turns left **WITH A WIDE FLAT END.**

#3 is the message: **"TRAVEL DOWN FROM THIS POINT TO A LARGE FLAT AREA, THEN TURN TO THE LEFT INTO A CONTINUING WIDE FLAT AREA OR CANYON/ ON YOUR WAY HOMEWARD"**

--- OR **"B"**:

#4 is a **V** shaped notch along the top line of rock #2.

#5 is a flat area made so that a person can sit and look through the notch #4 the same as you would look through the rear site of a rifle or place a spyglass in the notch.

#6 is a portion of the cliff face that has been cut away which now allows for sighting through notch #4 to see a **MAJOR OUT TRAIL MONUMENT** about a mile and a half away on a mountain slope.

#7 is the inverted **"V"** (refer to) which points to this trail **"B"** as being the best (and Royal) trail out.

"A"

A-1: IS OF COURSE A DIRECTIONAL POINTER. IT POINTS TO POINT "2" OF A COMPASS TYPE ROCK TRIANGLE. THE DARK AREA IS A PART OF A CAVE ENTRANCE THAT IS ABOUT EIGHT FEET DEEP. THE LARGE BOULDER TO THE RIGHT IS ABOUT 6 FEET OUT FROM THE CAVE ENTRANCE. NOTE THE 3 ROCKS ABOVE THE POINTER, ONE IS ALMOST WHITE, POSSIBLY MEANT TO BE AN "EYE-CATCHER".

A-2: TO THE RIGHT OF NUMBER 2 IS POINT "2" OF THE TRIANGLE, SEE COMPASS ROCK PAGE FOR PHOTO. ABOUT ONE HOUR EACH WAY OF "HIGH NOON" IS THE ONLY TIME THAT SUN/SHADOW SIGNS CAN BE SEEN.
THE CREW MEMBER IS HOLDING A MAGNETOMETER DURING OUR SURVEY.

"B"

B-1: IS AGAIN A POINTER, IN THIS INSTANCE THE LONG ARM IS CUT SO DEEP THAT IT CAN BE SEEN ALMOST ALL DAY LONG AND LOOKS LIKE A NORMAL CRACK OR CREVICE. BUT NOTE THE ROCK THAT PROJECTS OUT AT THE TOP OF THE LONG CURVED ARM. AT HIGH NOON THE SUN CAUSES A SHADOW TO FALL BELOW THE ROCK AND FORM THE POINTER. THE MESSAGE IS: "GO UP THE LARGE CREVICE, #2."

B-2: THIS WIDE CREVICE IS VERY STEEP, ABOUT 150 FEET UPWARD, AND I DON'T THINK ANYONE WOULD TRY TO CLIMB IT,--BUT IT'S EASY, IT HAS STEPS CUT INTO THE ROCK ALL THE WAY TO THE TOP, HOWEVER THE STEPS DON'T BEGIN AT THE BOTTOM, THEY BEGIN ABOUT 10 FEET UP, AND THE BRUSH COMPLETELY HIDES THIS TRAIL OF STEPS.

This is a very large and detailed trailhead monument. #1 is the trail directional pointer. It's approximately a 2 1/2 ton rock held up by volleyball sized rocks to allow light/sky to shine through. #2 is a rock with a 90° notch indicating that this trail is a single trail, no choices. #3 is an **INDIAN HEAD** with a top knot, a warning to be alert along the trail. #4 is the top knot and #5 is the **HOLE**, the mine/treasure symbol. This monument says: the trail into the mountains begins here. It is a single trail and if problems arise, turn around and come out the way you went in. **BE ALERT**. This trail will probably encounter **INDIANS** before you reach your goal, the **"HOLE"**.

Section 7:

DEATH TRAPS

Foreword

THE ONLY GOOD THING, OR RATHER, **POSITIVE ASPECT RELATED TO DISCOVERING A "DEATH TRAP" IS THE ABSOLUTE CONFIRMATION THAT YOU HAVE "BROKEN THE MAP'S CODE CORRECTLY" OR FOLLOWED THE INSTRUCTIONS PROPERLY —** FOR YOU ASSUREDLY ARE IN THE IMMEDIATE AREA OF A MAJOR MINE OR TREASURE, OR POSSIBLY BOTH. TUNNELS OF RICH MINES ARE KNOWN TO HAVE BEEN FILLED WITH TREASURES PRIOR TO THE COVERING AND CONCEALING OF THE MINE ITSELF.

BY FAR, THE MAJORITY OF ALL SPANISH DEATH TRAPS ARE IN THE IMMEDIATE "AREA" OF THE ENTRANCE TO THE MINE/TREASURE AND **NOT WITHIN THE MINE OR TREASURE ROOM ITSELF.**

SUBTLE AND MAN MADE CHANGES WERE MADE IN TOPOGRAPHY, MARKINGS INSCRIBED ON BOULDERS, LARGE ROCKS "ARRANGED", BOULDERS "RESHAPED", ETC., IN THE IMMEDIATE "ENTRANCE" AREA. THESE WERE SET-UP TO **"ENTICE/ACT AS A MAGNET", TO MISLEAD AND BEGUILE ANYONE WHO WAS CLEVER ENOUGH TO GET WITHIN A COUPLE OF HUNDRED FEET OF THE ACTUAL "ENTRANCE" AND DRAW HIM AWAY FROM THE "JACKPOT" AND TO ONE OF THE NEARBY DEATH TRAP LOCATIONS.**

STUDY THE DEATH TRAPS AND **COMMIT THE TRAP WARNING SIGNS AND SYMBOLS TO MEMORY.** YOUR KNOWLEDGE AND MEMORY MAY SAVE A LIFE, POSSIBLY YOUR OWN. WHAT COULD BE WORSE THAN SPENDING MANY YEARS OF SEARCHING, **AND THEN, WHEN WITHIN "A STONE'S THROW" OF YOUR GOAL —** YOU GET SUCKER PUNCHED.

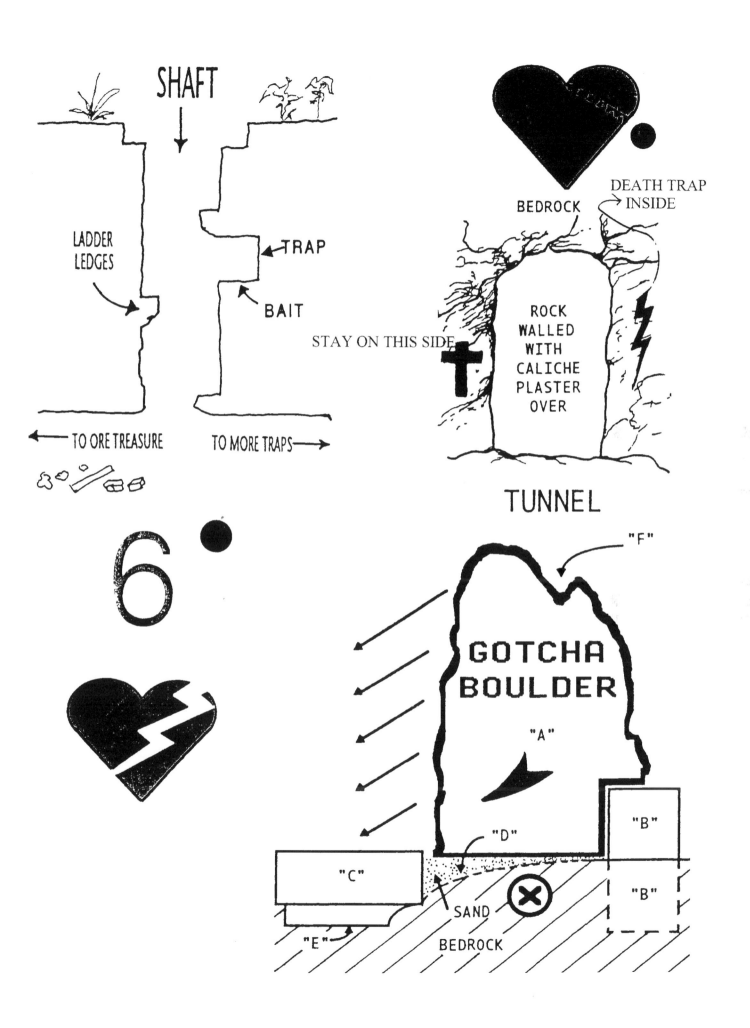

SHAFT

LADDER LEDGES

TRAP

BAIT

← TO ORE TREASURE

TO MORE TRAPS →

BEDROCK

DEATH TRAP INSIDE

STAY ON THIS SIDE

ROCK WALLED WITH CALICHE PLASTER OVER

6

TUNNEL

"F"

GOTCHA BOULDER

"A"

"D"

"B"

"B"

"C"

SAND

"E"

BEDROCK

THIS IS THE "DEATH TRAP GRANDE", THE MOST COMMON TRAP FREQUENTLY USED BY THE SPANISH. THIS TRAP WAS DESIGNED TO CRUSH EIGHT TO FIFTEEN SEARCHERS AT ONCE.

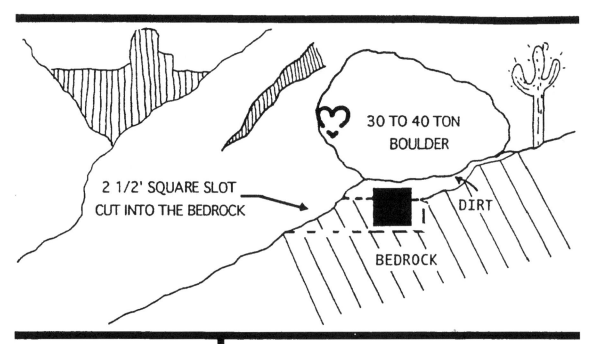

30 TO 40 TON BOULDER

2 1/2' SQUARE SLOT CUT INTO THE BEDROCK

DIRT

BEDROCK

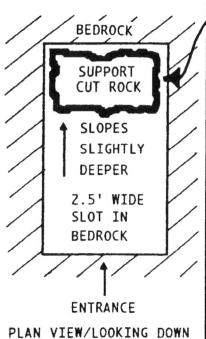

BEDROCK

SUPPORT CUT ROCK

SLOPES SLIGHTLY DEEPER

2.5' WIDE SLOT IN BEDROCK

ENTRANCE

PLAN VIEW/LOOKING DOWN

THIS ROCK WAS CUT TO FIT INTO THE "SLOT" VERY TIGHTLY AND PLACED AGAINST THE BACK WALL OF THE SLOT FOR FURTHER UPRIGHT SUPPORT AS WELL AS SUPPORT OF NOT MOVING OR FALLING "BACKWARDS" DURING A "TEMBLOR" (EARTHQUAKE) OR OTHER MAJOR LAND DISTURBANCE. THESE "SUPPORT ROCKS" **ALWAYS MUST PROJECT TWELVE (12) TO EIGHTEEN (18) INCHES ABOVE THE SIDES OF THE BEDROCK WALLS WITH THE BOULDER RESTING ON IT.**

DIRECTLY BELOW THE POINTER OF THE BROKEN HEART WILL BE THE "ENTRANCE" OR BEGINNING OF THE SLOT CUT INTO THE BEDROCK AND

SLOT CUT INTO THE BEDROCK AND USUALLY FOUND ONLY ABOUT A FOOT OR SO UNDER THE SURFACE OF THE GROUND ON THE SLOPE.

WHEN YOU FIND ONE OF THESE DON'T DO AS WE HAVE DONE AND CHIP AWAY AT THE SUPPORT ROCK THINKING IT IS JUST A "CAP ROCK" THAT SEALS THE ENTRANCE TO THE TREASURE ROOM OR MINE. IT SURELY DOESN'T LOOK LIKE A "TRAP"—IT LOOKS LIKE A MOST PERFECT PLACE AND WAY TO "HIDE/CONCEAL" AN ENTRANCE TO A MINE/TREASURE. ALSO, HOW COULD SUCH A SMALL 2-1/2 SQUARE FOOT ROCK HOLD UP A 30 to 40 TON BOULDER 25' WIDE BY 12' DEEP AND 9' TALL — AND AFTER ALL, DIDN'T THE BOULDER HAVE A HEART, THE SPANISH SYMBOL FOR GOLD, CARVED INTO ITS FACE?

THIS GRANDE TRAP IS ONE OF THE FULLY EXPOSED TRAPS THAT SITS IN TOTALLY INOCENT VIEW. WHEN THE BOULDER IS RELEASED THE WORKERS AND WATCHERS ON THE LOWER SIDE ARE IMMEDIATELY CRUSHED. ADDITIONALLY, AND ESPECIALLY IF THIS TRAP IS HIGH ON A MOUNTAINSIDE SLOPE, THIS BOULDER'S ROAR DOWN HILL WILL TEAR A PATH TWENTY FIVE FEET WIDE AND A FEW FEET DEEP POSSIBLY CATCHING A FEW MORE LAID BACK INTRUDERS.

EVEN IF THIS GRANDE BOULDER DOES NOT STRIKE ANOTHER INTRUDER ON IT'S RUSH DOWNHILL, THE NOISE IT MAKES AND THE PATH IT CUTS WILL SURELY PUT THE FEAR OF GOD — OR OF THE SPANISH, INTO THE HEARTS OF ANY UNTOUCHED INTRUDERS.

SPAIN'S COMMENTS ON THIS ADDED MENTAL IMPACT IS THAT "IT MAY WELL BE ENOUGH TO CAUSE THE REMAINING INTRUDERS TO ABANDON THEIR SEARCH AND RAPIDLY DEPART THE AREA SAYING A PRAYER OF THANKSGIVING TO THEIR PAGAN GODS."

NORMALLY I'M NOT SHOWING SIMILAR DEATH TRAPS IN THIS BOOK. HOWEVER, THIS ONE IS UNIQUE BECAUSE OF ITS **"BAIT OF GOLD"** AND POPULARITY WITH THE SPANISH. IT WAS USED IN **SHAFT WALL** CAVES FOR SHAFT PLUG TRAPS BUT INSTALLED MAINLY IN TUNNELS AND AT THE END OF PASSAGEWAYS.

THE WOODEN BOX MUST HAVE A LID AND HANDLES ON THE THREE EXPOSED SIDES. ALSO THE BOX SHOULD BE AT LEAST 3-1/2 FEET WIDE AND OF COURSE WIDER THAN THE BASE OF THE GOTCHA BOULDER.

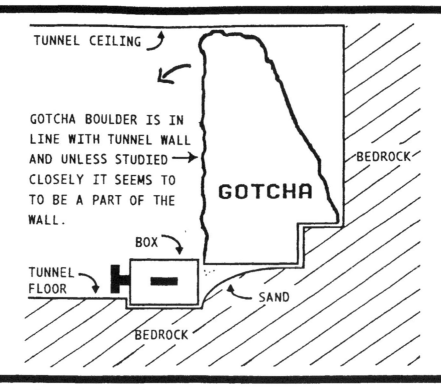

THE BOX IS SET IN A RECESSED AREA OF THE BEDROCK FLOOR. THE HANDLES ASSIST IN THE MOVING/PICKING-UP OF THE BOX WHICH HAS ABOUT 100 POUNDS OF GOLD ORE WITH MAYBE A SPECIMEN PIECE OR TWO IN READY VIEW.

UPON MOVING THE BOX, THE SAND THAT HAS BEEN THE PARTIAL BASE FOR THE GOTCHA BOULDER IN CONJUNCTION WITH THE BACK WALL OF THE BOX IS RELEASED AND THE GOTCHA TOPPLES DOWN.

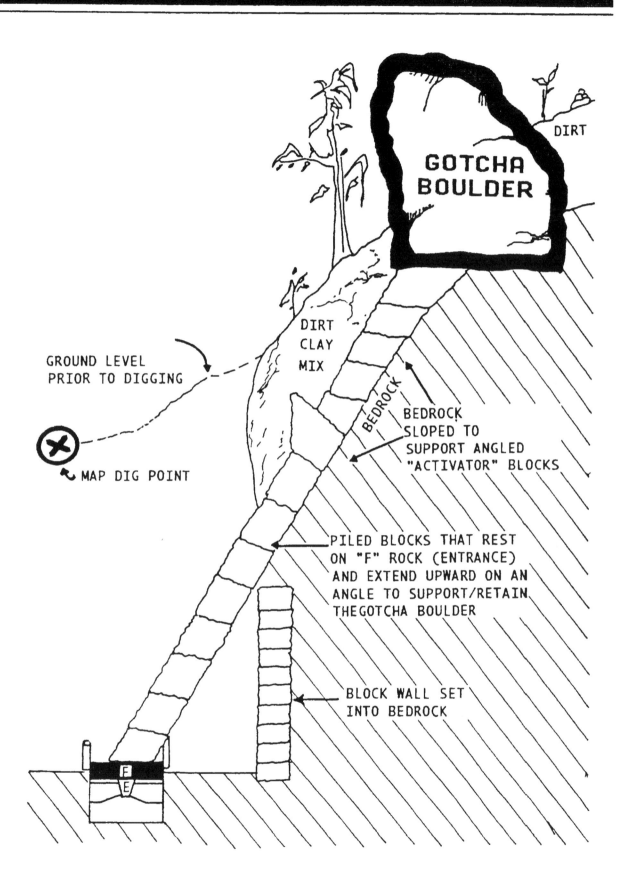

DIRT

GOTCHA
BOULDER

GROUND LEVEL
PRIOR TO DIGGING

DIRT
CLAY
MIX

BEDROCK

BEDROCK
SLOPED TO
SUPPORT ANGLED
"ACTIVATOR" BLOCKS

MAP DIG POINT

PILED BLOCKS THAT REST
ON "F" ROCK (ENTRANCE)
AND EXTEND UPWARD ON AN
ANGLE TO SUPPORT/RETAIN
THE GOTCHA BOULDER

BLOCK WALL SET
INTO BEDROCK

F
E

20069838R00069

Made in the USA
Charleston, SC
25 June 2013